W0225616

East-West Crosscurrents in Higher Education

Series editor

Ruth Hayhoe, University of Toronto, Toronto, ON, Canada

This book series focuses on higher education crosscurrents between Asia and the West, including traditional comprehensive universities, normal universities for teachers, higher vocational institutions, community colleges, distance and on-line universities and all the differing approaches to higher education emerging under processes of massification and diversification. It gives attention to the ways in which the Asian context shapes the internationalization of higher education and the response to globalization differently from that of the West, as well as new phenomena that are arising in the interface between these two broad regions, such as higher education hubs and regional networks of collaboration. Lastly, it will highlight the growing reciprocity between these two regions, whose higher education systems have grown from such deeply different historical roots.

Higher Education has deep roots in the cultures and civilizations of diverse regions of the world, but perhaps the most influential models shaping contemporary globalization come from Europe and China. Universities established in Europe in the Middle Ages have developed into what is now described as the "global research university," a model profoundly shaped by 19th century Germany and 20th century America, and spread around the world both through colonization and the emulation of its scientific achievements and contribution to nation building. A millennium earlier China spawned another influential model, characterized by close integration within a meritocratic bureaucracy that entrusted governance to those who could demonstrate their knowledge through written examinations. The Chinese model was greatly admired in Europe from the time it was introduced in the 16th century, and one can see its contours in what Burton Clark described as the "continental model" in contradistinction to the "Anglo-American model" epitomized in the global research university.

What has become clear in the maelstrom of globalization, which has stimulated the growth of a global knowledge economy and created circumstances where nations consider higher education as crucial to remaining competitive, is that the integration of core features from both models would be optimal: from Asia, a tradition of strong state support for and involvement in higher education, which is crucial for good governance and social advancement; and from Europe and North America, the ideas of university autonomy and academic freedom, which are essential to promoting scientific creativity and innovation.

More information about this series at http://www.springer.com/series/13844

Jiabin Zhu

Understanding Chinese Engineering Doctoral Students in U.S. Institutions

A Personal Epistemology Perspective

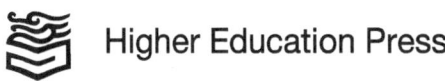 Higher Education Press

 Springer

Jiabin Zhu
Graduate School of Education
Shanghai Jiao Tong University
Shanghai
China

ISSN 2364-6810   ISSN 2364-6829   (electronic)
East-West Crosscurrents in Higher Education
ISBN 978-981-10-1135-1   ISBN 978-981-10-1136-8   (eBook)
DOI 10.1007/978-981-10-1136-8

Jointly published with Higher Education Press

Library of Congress Control Number: 2016942901

Printed on acid-free paper

This Springer imprint is published by Springer Nature
The registered company is Springer Science+Business Media Singapore Pte Ltd.

Foreword

Over the past three decades, there has been a continuing flow of Chinese students to Western countries—in particular, the United States—for higher education. The learning experiences of overseas Chinese students in the US higher educational system and their perspectives on such experiences say much about East–West crosscurrents in higher education. How do these Chinese students perceive the US higher educational system? How would they compare the Chinese and US higher educational systems? This volume constructs a cogent dialogue between the US and Chinese higher educational systems through the lens of personal epistemology.

In the realm of personal epistemology, another interesting East-West dialogue has been formulated by the author's testing of Perry's scheme of intellectual and ethical development, rooted in epistemology, among Chinese students studying in the United States. Given that it was originally established on the basis of an investigation into the intellectual development of white male students at Harvard University, can the Perry scheme be effectively applied to overseas Chinese students? The research participants described in this book—Chinese engineering students studying in top U.S. research universities—represent an elite group who had studied in top-tier 211 and 985 higher educational institutions in Mainland China. Dr. Zhu's study offers a unique comparison between the overarching profile of the epistemological development of a group of overseas Chinese students with that of the original elite white male students in Perry's study, which in itself reflects another dimension of East–West crosscurrents in higher education at both the theoretical and the practical level.

Readers will learn about the fascinating experiences of Chinese students as they wrestled with academic decisions and life decisions, the challenges they faced and the struggles they went through, as well as their endeavors and their growth—both academically and personally, especially in the domain of cognitive development. Chinese graduate students represent one of the largest and fastest-growing international student bodies in the US. A deep understanding of their learning experiences will shed valuable light on the ongoing East-West conversation. For example, how has their epistemological thinking been shaped by the rigorous engineering graduate programs, by the individualistic yet cooperative campus culture, and, further, by the larger context of American culture and society? What personal, professional, or cultural factors have proved to be relevant or even useful in promoting these students' cognitive development?

By using a predominantly quantitative-driven, mixed research method design, with a questionnaire survey supplemented by interviews, Dr. Zhu has provided insightful answers to the above and many other important questions. She has succeeded in telling a holistic and in-depth story of the cognitive development of overseas Chinese students in the United States. Readers will find the story told by Dr. Zhu, a young Chinese scholar who went through both the Chinese and US higher educational systems herself, engaging and thought-provoking.

Li-fang Zhang
The University of Hong Kong

Preface

The US has experienced a large surge of foreign talent, as evidenced by the large number of international students enrolling each year in the science and engineering fields. Among the foreign countries and economies, China ranks top in the number of doctorate degree recipients from US institutions in the science and engineering fields. This book focuses on studying the epistemological development of Chinese engineering doctoral students who are pursuing degrees in US institutions.

This work stems from my personal interest on the Chinese student population in US and my experiences as one of them through seven years of study in the US. It is certainly not a one-man job. First of all, I want to give thanks to my heavenly Father who sustained me through this work. I would like to give special thanks to my advisor Dr. Monica F. Cox. Her professional guidance and inspirations have encouraged me throughout my doctoral training. She and her husband, Ishbah Cox, have showered wonderful blessings on me and served as role models for me. Deep appreciation goes to professors at the School of Engineering Education and other departments at Purdue University. They offered critical insights, helped me, and worked with me through many big and small steps. They are Dr. Brent Jesiek, Dr. Yating Haller, Dr. William Graziano, Dr. Phillip Wankat, Dr. David Radcliffe, Dr. Ruth Streveler, Dr. Robin Adams, Dr. Alice Pawley, Dr. Donna Ennerson, and many other professors.

Much gratitude goes to staff members at the School of Engineering Education at Purdue University. Their kindness and helpful spirits have been of great encouragement and inspiration to me. They are Ms. Loretta Mckinniss, Ms. Tamara Hare, Ms. Cindey Hays, and many others. I would like to also acknowledge the professional guidance from Ms. Cindy Lynch. Also, I want to thank Mr. Joe J.J. Lin for his kind help and valuable feedback on this research. I want to acknowledge Dr. Lifang Zhang who kindly supported this research by sharing the copy of Zhang's Cognitive Development Inventory. Also, I want to thank all of the participants in this study. It is their help that made this study possible.

Many thanks go to colleagues and classmates at Purdue University and around the world. They include:

Ms. Jeremi London, Mr. Benjamin Ahn, Dr. Rocia Chavela, Dr. Osman Cekic, Dr. Nathan McNeill, Mr. Mark Carnes, Ms. Ming-chien Hsu, and many others.

Still more thanks go out to my friends, brothers and sisters in Christ, who have been loving and supportive throughout my years at Purdue University, West Lafayette, IN, USA.

Last but not least, special thanks to my dear mother, who showered me with sacrificial love and kindness. This work would have been impossible without her consistent support.

Shanghai, China   Jiabin Zhu

Contents

About the Author

Jiabin Zhu is currently an assistant professor at the Graduate School of Education, Shanghai Jiao Tong University. She received her Bachelor of Science degree in Physics from East China Normal University and her Master of Science degree in Optics from the Chinese Academy of Sciences. She received a second Master's Degree in Biomedical Engineering and a Ph.D. in engineering education from Purdue University, West Lafayette, IN, USA. She was the awarded the 2012 Bilsland Dissertation Fellowship Award and 2012 Bilsland Strategic Initiatives Fellowship Award. She received the Best Graduate Student Paper Award from the Graduate Studies Division in the 2012 American Society for Engineering Education (ASEE) annual conference. She also received the Outstanding Research Award from the College of Engineering, Purdue University. For this work, she was awarded the 2013 Dissertation Award from the School of Engineering Education, Purdue University.

Her primary research interests relate to the epistemological development of college and graduate students, comparative study methods and frameworks in engineering education, global engineering, and mentoring of engineering graduate students. She is currently in charge of multiple research projects sponsored by the Chinese Ministry of Education and Shanghai Municipal Government.

Abbreviations

C	Commitment within Relativism
D	Dualism
EBI	Epistemological Beliefs Inventory
EBS	Epistemological Belief Survey
EQ	Epistemological Questionnaire
KCM	Knowledge Construction and Modification
M	Multiplicity
MER	Measure of Epistemological Reflection
R	Relativism
RCI	Reasoning about Current Issues Test
ZCDI	Zhang's Cognitive Development Inventory

Chapter 1
Influx of International Talents in the United States: An Introduction

1.1 Influx of International Talents to the US

The rapid globalization of the economy and technology over recent decades has led to an increasing mobility of students all over the world. Currently, the US is engaged in a competition to attract the "world's best and brightest" students and scholars (NAFSA 2006, p. 6). The US has seen a surge of international talents, with a high of 819,644 students enrolled in its higher educational institutions in the 2012/2013 academic year (Institute of International Education, IIE 2013). China ranked as the top place of origin and accounted for over a quarter of the international students. Finn (2010) found that, among the foreign-born PhD recipients in science and engineering, 62 % of them stayed in the US for five years (2003–2007) after their graduation in 2002.

Among those foreign countries or regions from which international students come, China has ranked as the top place of origin in 2009/2010, providing 18.5 % of all international students in the US (IIE 2010). The increase of Chinese students in US institutions reached an all-time high in 2009/2010 with 127,628 students, a 29.9 % increase from 2008/2009. Among these Chinese students, 52.1 % of them engaged in graduate level studies and 31.3 % in undergraduate studies (IIE 2010). For science and engineering fields, China also ranks highest in the top ten countries that had the most science and engineering doctorate degree recipients (32,973) from US institutions (Fig. 1.1) (NSF 2010a). Chinese science and engineering PhD students also are enrolled in larger numbers at US institutions than are students from any other foreign country (Fig. 1.2) (NSF 2010b). Moreover, Finn (2010) noted 92 % of doctorate recipients from China stayed five years after graduation in 2007. This stay rate was higher than for students from any other country in the US.

According to Chinese data sources (MYCOS 2010), more Chinese students who travel abroad to pursue graduate degrees graduate from 211 type universities (i.e., around 100 universities designated in a national "211 project" by the Chinese Ministry of Education in 1995 to receive support from the Chinese government to

© Springer Science+Business Media Singapore and Higher Education Press 2017   1
J. Zhu, *Understanding Chinese Engineering Doctoral Students in U.S. Institutions*,
East-West Crosscurrents in Higher Education, DOI 10.1007/978-981-10-1136-8_1

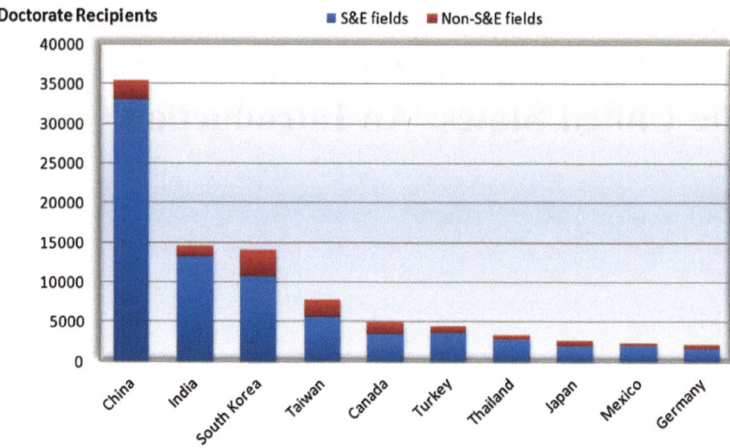

Fig. 1.1 Top ten counties/economies of foreign citizenship for US doctorate recipients: total, 1999–2009 (NSF 2010a). *Note* China includes Hong Kong

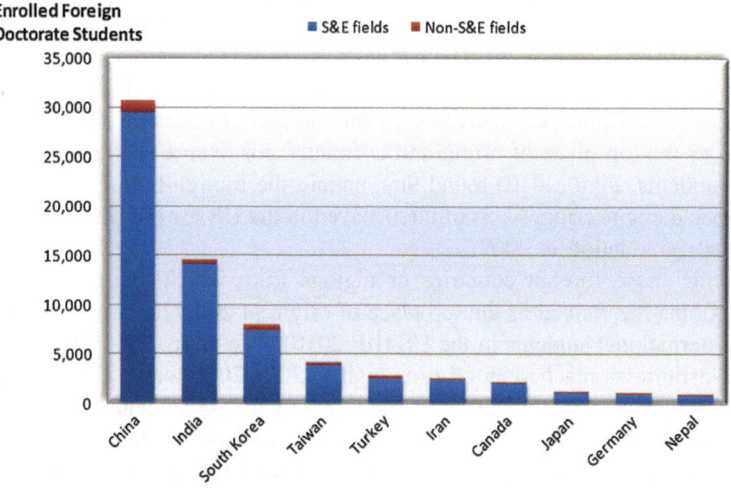

Fig. 1.2 Top ten counties/economies of foreign citizenship for US doctorate enrollment: data for 2009 (NSF 2010b)

facilitate said universities' development in educational quality, research, and administration; see archives from the Chinese Ministry of Education website 2009) than students at non-211 type universities (Fig. 1.3). This indicates that the majority of the 1–2 % of Chinese students migrating to the US to pursue science and engineering degrees were primarily high achievers from 211 type universities.

The flow of Chinese science and engineering students to US institutions has increased dramatically for several reasons:

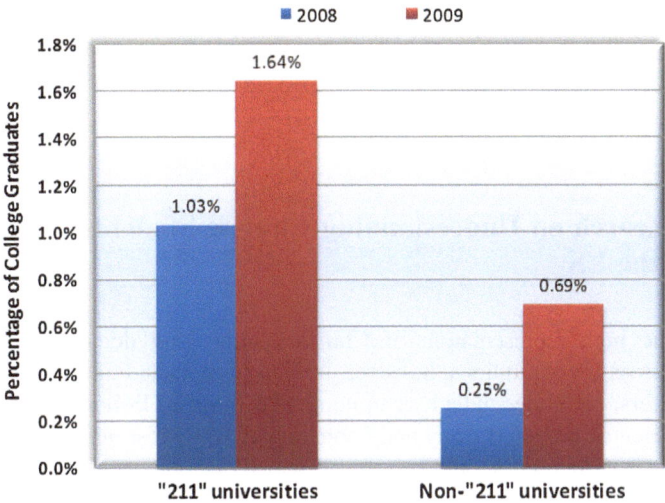

Fig. 1.3 Percentage of college graduates who studied abroad from 211 and non-211 Chinese universities (MYCOS 2010, p. 60)

First, China's Reform and Opening policy in 1978 and an important talk on 23 June 1978 by the then Chinese Vice Chairman Deng on increasing the number of students studying abroad, with a focus on natural science disciplines, set a historical background for the flow of Chinese students into foreign counties (see archives from the Chinese Government website 2009). Also on this historical background was the normalization of diplomatic relations between China and the US after President Nixon's February 1972 visit to China (see archives from the Embassy of the United States website 2013), which set the backdrop for the flow of Chinese students to US institutions.

Second, a policy about largely expanding undergraduate student enrollment was issued in Chinese higher education systems in 1999 to increase domestic talents and to extend the Chinese domestic economic need after the 1997 Asian economic crisis (Yang 2011). This policy led to a 33.5 % annual average increase in undergraduate student enrollment between 1999 and 2001 and a 24.8 % annual average increase between 1999 and 2004 (Yang 2011) in China. The increase in undergraduate student enrollment enlarged the potential Chinese student pool that would possibly pursue study abroad opportunities.

Third, several major national projects were launched to focus on enhancing educational quality, academic and research standards in Chinese higher education, such as "211 Project" (1995) and "985 Project" (1998) (see archives from the Chinese Ministry of Education website 2008). The implementation of these educational projects had endowed rich educational resources to key universities in China, which facilitated the production of more quality students in key universities, especially students in science and engineering disciplines.

All of the above-mentioned and other economic and socio-political factors, directly or indirectly, increase the representation of Chinese students in US institutions, especially the number of students in the science and engineering disciplines.

1.2   Research on Understanding Chinese Students in the US

Despite the high representation of Chinese engineering doctoral students and scholars across US institutions, however, foreign-born, including Chinese scientists and engineers, are understudied when compared to their US-born peers and other underrepresented groups (Corley and Sabharwal 2007). Most *quantitative research* in this area has focused on the research productivity, job satisfaction, and career trajectory issues of foreign-born scientists and engineers (Marvasti 2005; Wells et al. 2007; Corley and Sabharwal 2007; Mamiseishvili and Rosser 2009; Lin et al. 2009). For example, Corley and Sabharwal (2007), using the 2001 Survey of Doctorate Recipients data gathered from the NSF, compared productivity, work satisfaction, and career trajectories of foreign-born scientists and their US peers. They concluded that foreign-born scientists had a higher demonstrated level of productivity (measured by published articles, books, papers, and patent activity) and lower salaries and work satisfaction levels than did their US peers. Using the structural equation modeling of 2004 National Study of Postsecondary Faculty data, Mamiseishvili and Rosser (2009) showed that international faculty members were significantly more productive in research but were less productive in their teaching and service than were their US-born peers.

Among the limited *qualitative research* on foreign-born scholars, major efforts have been spent primarily on the exploration of their adjustment issues, such as experiencing a sense of isolation, balancing family life and career, experiencing a lack of collegiality, and overcoming language barriers (Ye 1992; Seagren and Wang 1994; Thomas and Johnson 2004; Skachkova 2007; Wang 2009). Very little attention has been devoted to studying the intellectual development or the epistemological development of these groups under the US system, although epistemological development studies among their US peers have been going on for nearly four decades.

In the realm of *epistemological development research*, significant efforts have been devoted to understanding young adults in their epistemological development through the higher education system, especially at the collegiate level, since the pioneering work by William Perry in the 1960s (Perry 1970; Belenky et al. 1986; Baxter Magolda 1992; King and Kitchener 1994; Zhang 1995, 2002). Perry described a nine-position theory of the college student's epistemological development from a dualistic to a constructive view. Perry's nine-position theory can be summarized into four main developmental stages: *Dualism* (*D*), *Multiplicity* (*M*),

Relativism (R), and *Commitment within Relativism (C)* (Culver and Hackos 1982). In the context of engineering education, researchers point out that engineering college students have shown slow development according to Perry's theory because of differing education methods and the context of engineering as compared to liberal arts (Wankat and Oreovicz 1993). Often, liberal arts students are given more opportunities to develop multiple perspectives and allowed greater tolerance towards ambiguity, which potentially facilitates this development. It is speculated that graduate students will develop further in this transition to a "constructive knower," which was indicated by a study on liberal arts students (Kitchener and King 1981); however, no substantial evidence has been provided to support this claim among engineering students.

Perry (1970) and some later researchers, such as Belenky et al. (1986), Baxter Magolda (1992), and King and Kitchener (1994), based their research and findings mostly on US institutions and US populations. Recently, epistemological development research has also been performed among Chinese college students from Chinese universities. Starting from 1995, Perry's theory has been applied to study US and Chinese college students by Zhang (1995, 2002). However, students in Zhang (1995, 2002, 2004) series of studies represented a variety of areas (e.g., education, liberal arts, science, sociology, etc.). Despite these studies, there is still scarce information available exploring the epistemological development of graduate-level engineering Chinese students. Considering the prominent representation of Chinese students in doctoral engineering education, a significant expectation of students to develop cognitively in higher education, and current research focusing largely on Chinese students' adjustment and socialization and not on cognitive development, this research attempts to examine the epistemological development of Chinese engineering doctoral students in US institutions framed within the context of Perry's theory.

1.3 Overview of This Work

This work starts by setting a theoretical and methodological foundation through landing the topic of personal epistemology on the larger scale of cognitive psychology. In Part I, Chaps. 2, 3 and 4, this work provides a review of current epistemological developmental theories and introduces the applications of theories and current measurement tools in understanding students' epistemological development.

Chapter 2 starts with a conceptualization of personal epistemology, i.e., asking where personal epistemology is situated in a bigger picture of cognition or the cognitive process. Next, it reviews current cognitive developmental theories in personal epistemology, including Perry's theory and its extensions over the past 40 years. A synthesis of these theories suggests that five models all describe young adults' epistemological development through a common trend from a dualistic view of knowledge to a contextual, constructivist perspective, which was used in this work.

Throughout the past 40 years of research in the cognitive developmental field, multiple research methods, both qualitative and quantitative, were used in exploring young adults' epistemological development. Chapter 3 provides an overview of available qualitative methods and quantitative tools in exploring and measuring young adults' epistemological development.

Built upon the discussion about the theories and measurement tools for personal epistemology, Chap. 4 introduces their current applications in addressing students' personal epistemology. In particular, I focus on integrating present research findings that are related to engineering students, doctoral students, and Chinese students, respectively. This chapter provides an overview of research that is highly relevant to the effort to understand Chinese engineering doctoral students' epistemological development.

Built on the theoretical and methodological foundation, Part II, Chaps. 5 and 6, addresses the core methodological approach that was adopted in this research.

As a critical step for mapping the epistemological development profile of Chinese engineering students on a large scale, Chap. 5 introduces the construction and validation of a quantitative survey built upon prior measurement tools in the context of Perry's theory. In particular, details were provided on the content validation and structural validation process of this tool.

Using the survey developed in Chap. 5, Chap. 6 describes the motivation and the process of an explanatory mixed-method research design. It starts by explaining the dual purposes for an explanatory mixed-method design, in which quantitative and qualitative data were collected sequentially. As Creswell stated, this methodological design captures "the best of both quantitative and qualitative data—to obtain quantitative results from a population in the first place" and "refine or elaborate these findings through an in-depth qualitative exploration in the second phase."

The main sections of this book, Parts III and IV, are composed of the findings from this study. Chapters 7 and 8 of Part III present quantitative profiles of Chinese engineering doctoral students' epistemological development using the survey described in Chap. 5.

In Chap. 7, the prominent epistemological development stages of Chinese engineering doctoral students from five Midwestern universities will be provided to give an overview of their epistemological development. The majority of Chinese engineering doctoral students have demonstrated a mature level of epistemological thinking in the context of Perry's theory. Trends and findings related to the overall profiles will be discussed in detail.

In Chap. 8, I highlight factors related to students' epistemological development. This is done by examining differences that are present in survey subscales concerning various demographic parameters. Such factors include the varied students' current academic progress within doctoral studies, their prior master's education, places of origin, and their enrolled universities. The results point to the possible effect of US engineering graduate training and demographic factors on students' epistemological development.

Part IV, which includes Chaps. 9 and 10, presents a qualitative exploration of Chinese engineering doctoral students' epistemological thinking styles through an investigation of students' thinking patterns and behavioral characteristics at different epistemological developmental stages and factors that were associated with their development of mature epistemic thinking.

Built upon findings from the quantitative survey results, Chap. 9 examines Chinese engineering doctoral students' epistemological thinking in a qualitative manner using Perry's theory. Based on one-on-one interviews, this chapter presents practical instances that represent students' epistemological thinking from their experiences in US doctoral programs. Rich details related to students' thinking processes, thinking styles, and behavioral patterns are summarized to offer detailed descriptions for each epistemological developmental stage, thereby presenting stories of each epistemological thinking.

Chapter 10 explores factors associated with students' epistemological development to advanced thinking. These factors include impacts from personal aspects (e.g., advisors and lab mates), experiential influences (e.g., group discussions, failures, pressures, or obstacles, and experiences with open-ended projects) and contextual dimensions (e.g., being in the US or on a US campus). Sample quotes that illustrate the impact from these factors were incorporated to demonstrate the varied influence on students' epistemological thinking.

References

Archives from Chinese Government website. (2009). Retrieved from http://www.gov.cn/test/2009-09/30/content_1430681.htm (in Chinese).

Archives from Chinese Ministry of Education. (2008). Retrieved from http://www.moe.gov.cn/publicfiles/business/htmlfiles/moe/moe_2442/200810/39607.html

Archives from Chinese Ministry of Education website. (2009). Retrieved from http://www.moe.gov.cn/publicfiles/business/htmlfiles/moe/moe_2921/200909/52111.html (in Chinese).

Archives from Embassy of the United States website. (2013). Retrieved from http://beijing.usembassy-china.org.cn/highlevel.html

Baxter Magolda, M. B. (1992). *Knowing and reasoning in college.* San Francisco: Jossey-Bass.

Belenky, M. F., Clinchy, B. M., Goldberger, N. R., & Tarule, J. M. (1986). *Women's ways of knowing: the development of self, voice and mind.* New York: Basic Books.

Corley, E. A., & Sabharwal, M. (2007). Foreign-born academic scientists and engineers: Producing more and getting less than their U.S.-born peers? *Research in Higher Education, 48* (8), 909–940.

Culver, R. S., & Hackos, J. T. (1982). Perry's model of intellectual development. *Engineering Education, 72,* 221–226.

Finn, M. G. (2010). *Stay rates of foreign doctorate recipients from U.S. universities, 2007.* Oak Ridge Institute for Science and Education.

Institute of International Education, IIE. (2010). *Open doors 2010: Report on international educational exchange.* New York: Institute of International Education.

Institute of International Education, IIE. (2013). *Open doors 2013: Report on international educational exchange.* New York: Institute of International Education.

King, P. M., & Kitchener, K. S. (1994). *Developing reflective judgment: Understanding and promoting intellectual growth and critical thinking in adolescents and adults.* San Francisco: Jossey-Bass.

Kitchener, K. S., & King, P. M. (1981). Reflective judgment: Concepts of justification and their relationship to age and education. *Journal of Applied Developmental Psychology, 2*, 89–116.

Lin, Z., Pearce, R., & Wang, W. (2009). Imported talents: Demographic characteristics, achievement and job satisfaction of foreign born full time faculty in four-year American colleges. *Higher Education, 57*(6), 703–721.

Mamiseishvili, K., & Rosser, V. (2009). International and citizen faculty in the United States: An examination of their productivity at research universities. *Research in Higher Education, 51*(1), 88–107.

Marvasti, A. (2005). U.S. academic institutions and perceived effectiveness of foreign born faculty. *Journal of Economic Issues, 39*(1), 151–176.

MYCOS, My China Occupational Skills. (2010). 2010 Report on Chinese University students' employment.

NAFSA: The Association of International Educators. (2006). Restoring U.S. competitiveness for international students and scholars. Retrieved from NAFSA website: http://www.nafsa.org/resourcelibrary/default.aspx?id=9169

NSF, National Science Foundation. (2010a). Doctorate recipients from U.S. universities. Retrieved from http://www.nsf.gov/statistics/nsf11306/index.cfm

NSF. (2010b). Foreign science and engineering students in the United States. Retrieved from http://www.nsf.gov/statistics/infbrief/nsf10324/

Perry, W. G. (1970). *Forms of intellectual and ethical development in the college years: A scheme.* New York: Holt, Rinehart and Winston.

Seagren, A. T., & Wang, H. (1994). *Marginal men on an American campus: A case of Chinese faculty.* In Paper presented at The Annual Meeting of the Association for the Study of Higher Education, Tucson, Arizona.

Skachkova, P. (2007). Academic careers of immigrant women professors in the U.S. *Higher Education, 53*(6), 697–738.

Thomas, J. M., & Johnson, B. J. (2004). Perspectives of international faculty members: Their experiences and stories. *Education and Society, 22*(3), 47–64.

Wang, W. (2009). Chinese international students' cross-cultural adjustment in the U.S.: The roles of acculturation strategies, self-construals, perceived cultural distance, and english self-confidence. Retrieved from ProQuest. Austin: The University of Texas.

Wankat, P., & Oreovicz, F. S. (1993). Models of cognitive development: Piaget and Perry. In *Teaching engineering.* New York: McGraw-Hill. Retrieved from https://engineering.purdue.edu/ChE/AboutUs/Publications/TeachingEng/chapter14.pdf

Wells, R., Seifert, T., Park, S., Reed, E., & Umbach, P. D. (2007). Job satisfaction of international faculty in U.S. higher education. *Journal of the Professoriate, 2*(1), 5–32.

Yang, D. (2011). On "Chinese mode" in higher education. In *Jiang Su higher education* (Vol. 1, pp. 5–8) (in Chinese).

Ye, Y. (1992). Chinese students' needs and adjustment problems in a U.S. university. Retrieved from ProQuest. Lincoln: The University of Nebraska

Zhang, L. F. (1995). *The construction of a Chinese language cognitive development inventory and its use in a cross-cultural study of the Perry scheme.* Retrieved from ProQuest, The University of Iowa.

Zhang, L. F. (2002). Thinking styles and cognitive development. *The Journal of Genetic Psychology, 163*(2), 179–195.

Zhang, L. F. (2004). The Perry scheme: Across cultures, across approaches to the study of human psychology. *Journal of Adult Development, 11*(2), 123–138.

Part I
Epistemological Developmental Theories and Their Applications

Chapter 2
Epistemological Developmental Theories

This chapter provides an overview of current epistemological developmental theories. It starts with a conceptualization of personal epistemology, i.e., asking where personal epistemology is situated in a bigger picture of cognition or the cognitive process. Next, it reviews current cognitive developmental theories in personal epistemology, including Perry's theory and its extensions over the past 40 years. A synthesis of these theories leads to modifications of Perry's theory, which is used in this work.

2.1 Conceptualization of Personal Epistemology

Personal epistemology can trace its origin back to Piaget (1950, 1972) when he described his theory of intellectual development of children (referred to as *generic epistemology*). His theory on intellectual development initiated the interest of psychologists and educators in cognitive developmental theories. A central theme in his theory is a developmental progress through different stages throughout childhood and early adolescence. This central theme of a developmental progress in relation to knowing and knowledge is followed by many of the current models of epistemological development (Hofer and Pintrich 1997).

Before I review current models and theories in personal epistemology, I would like to first situate the development that will be discussed in this work within the broader territory of cognition. As shown in Table 2.1, Kitchener (1983) and Kuhn (2000) both distinguished different levels of cognitive processing.

Personal epistemology refers to what Kitchener (1983) defined as the third level of cognition, epistemic cognition (the first two levels are cognition and metacognition). It refers to reflections on "the limits of knowledge," "the certainty of knowledge," and "the criteria for knowing" (Kitchener 1983, p. 222).

In Kitchener's discussion about cognitive processing and dealing with ill-structured problems, he proposed a three-level model including *Cognition,*

© Springer Science+Business Media Singapore and Higher Education Press 2017
J. Zhu, *Understanding Chinese Engineering Doctoral Students in U.S. Institutions,*
East-West Crosscurrents in Higher Education, DOI 10.1007/978-981-10-1136-8_2

Table 2.1 Locating epistemological thinking in cognitive processing (Hofer 2001, p. 364)

3-level model of cognitive processing (Kitchener 1983)	3-level model of meta-knowing (Kuhn 2000)
Cognition	
Metacognition	Metacognitive knowing Metastrategic knowing 　Metatask knowledge 　Metastrategic knowledge
Epistemic cognition	Epistemological knowing

Metacognition, and *Epistemic Cognition*. At the first level, *Cognition* refers to an individual's ability to read, memorize, compute, etc. *Metacognition* has to do with the monitoring of the first level processes; *Epistemic Cognition* is related to reflections on "the limits of knowledge," "the certainty of knowledge," and "the criteria for knowing" (Kitchener 1983, p. 222). Prior findings (Flavell 1979; Kitchener 1983) suggest that cognitive and metacognitive processes emerge in young children and remain active throughout their lifespans, whereas *Epistemic Cognition* begins to develop in late adolescence and continues to shift in the adult years. In the context of solving ill-structured problems, Kitchener (1983) stated that, while *Metacognition* allowed one to choose different cognitive strategies for the purpose of tackling a specific task, *Epistemic Cognition* allows one to "interpret the nature of a problem and to define the limits of any strategy to solving it" (p. 226). *Epistemic Cognition* provides the foundation for adults by which they may deal with conflicting ideas in issues like logic, ethical choice, or career choice.

Besides the first level cognition, Kuhn (1999, 2000) introduced two sub-categories under the second level *Metacognition* (*Metacognitive Knowing* and *Metastrategic Knowing*) and a third level that parallels Kitchener's term, *Epistemic Cognition*, called *Epistemological Knowing*. *Metacognitive Knowing* refers to declarative knowing (knowing that), whereas *Metastrategic Knowing* refers to procedural knowing (knowing how). *Epistemological Knowing* develops when there is a transition from "simply knowing something is true to evaluating whether it might be" (Kuhn 1999, p. 22). As Hofer (2001) pointed out, the changes of criteria that we use to evaluate whether "something is true" are key areas of epistemological development.

Thus far, I have described the conceptualization of personal epistemology. In the rest of this document, cognitive development is only used to refer to epistemological development unless otherwise specified. Specifically, in the rest of this chapter, different frameworks related to epistemological development for young adults will be described. Although these researchers have used different names to refer to epistemological development, for example, "intellectual and ethical development" (Perry 1970) or "ways of knowing" (Belenky et al. 1986), their conceptualizations still fall within the scope of epistemological knowing itself. Therefore, terms like "intellectual development," if used separately (i.e., not in the case of "intellectual development and ethical development"), can be understood as interchangeable with epistemological development in most cases unless otherwise specified. The original terms of the authors are retained in this literature review to help understand their original ideas.

2.2   Perry's Theory

It has been four decades since the first issue of William Perry's work, *"Forms of Intellectual and Ethical Development: In the College Years"* (1970). Perry's theory delineates the cross sections of nine different "positions" along which development takes place. Following the conventions of the ongoing refinement of Perry's model over the past four decades (Knefelkamp 1974; Knefelkamp and Slepitza 1978; Moore 1991, 1994, 2002), the sequences of the nine positions proposed by Perry can be grouped into four major categories: *Dualism* (Positions 1 and 2), *Multiplicity* (Positions 3 and 4), *Relativism* (Positions 5 and 6), and *Commitment (within Relativism)* (Positions 7 through 9), as shown in Fig. 2.1. From Positions 1 and 2 to Positions 3 and 4, a person modifies a view of dualistic absolutism (right-wrong) to make room for simple pluralism, or so-called *Multiplicity*. From Positions 3 and 4 to Position 5, a person changes from the "simple pluralism of Multiplicity" into "Contextual Relativism," and then comes to Position 6, in which that person foresees the necessity of positioning him- or herself with some form of personal *Commitment* (as opposed to unquestioned commitment to simple belief) to in a relativistic world. In positions 7, 8, and 9, a person experiences a development of personal commitment. Positions 1 and 9 were extrapolations from Positions 2–8. They were added to make a full picture of intellectual and ethical development as noted by Perry (1970), although they were not observed in the original data.

Overview of Positions

Dualism
Position 1 Basic Duality. A person in this position perceives any knowledge, act, or value to be either "right" or "wrong." The world is divided into Authority[1]-right-we and Illegitimate-wrong-other. Any knowledge, act, or value that differs from Authority's world will be associated with error or evil, leaving the person with no alternative or vantage point to observe differently. Authorities cannot be separated from the Absolute.[2] Authorities are mediators of the right answers to any question related to knowledge, act or value.

---

[1] According to Perry's original description, Authority was defined as, "The possessors of the right answers in the Absolute" or "Pretenders to the right answers in the Absolute" (1970, Glossary page).

[2] Absolute was defined by Perry as, "The established Order; The Truth, conceived to be the creation and possession of the Deity, or simply to exist, as in a Platonic world of its own; The ultimate Criterion, in respect to which all propositions and acts are either right or wrong" (1970, Glossary page).

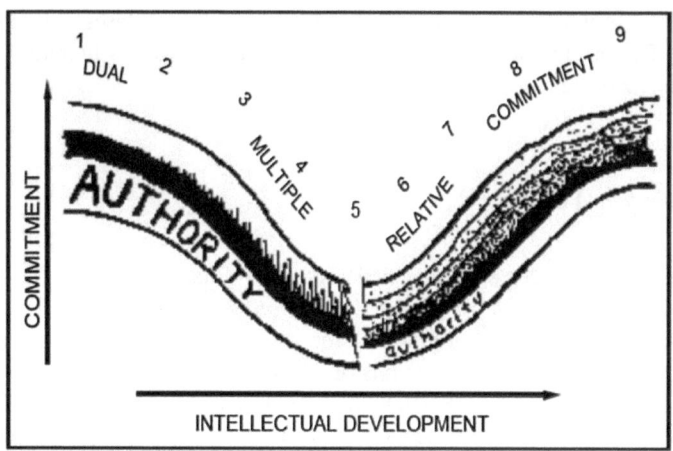

Fig. 2.1 Perry's model of intellectual development (Culver and Hackos 1982, p. 223)

Position 2 Multiplicity[3] Pre-legitimate. In this position, diversity in opinion and uncertainty has been given a place. They are viewed as an unwarranted impediment by "poor qualified Authorities" (Perry 1970, p. 9) to the right answer or a narrow area of freedom set by the Authority for student's own exploration. *Multiplicity* has not yet gained legitimacy in the Absolute.

Multiplicity
Position 3 Multiplicity Subordinate. A person in this position accepts uncertainty and diversity as unavoidable and legitimate but only in areas where Authority hasn't attained the right answers. Coupled with this accepted uncertainty is the student's puzzling about the standards against which Authority grades one's work in an educational setting when Authority has no correct answers him- or herself. In this position, he or she supposes that Authority grades based on nothing but "good expression."

Position 4. In developing the structure of this position, two alternative views have emerged from the students that Perry considered equivalent because they each represent an "ultimate extension or accommodation" of the old dualism system. (1) *Multiplicity Correlate.* In this position, *Multiplicity* is raised from a subordinate to a legitimate status, which is separate and equal to that of the Authority. An unstructured epistemological law with equal absolutism—"Everyone has a right to his own opinion" (p. 97) dominates the *Multiplicity* realm, while in the Authority's

---

[3]Multiplicity: "A plurality of 'answers,' points of view, or evaluations with respect to similar topics…with the implication that no judgments among opinions can be made" (Perry 1970, Glossary page). Compare Relativism.

domain there remains a right-wrong dualism. (2) *Relativism*[4] *Subordinate*. A person in this position perceives relativistic reasoning as what Authority wants. The weighing of different approaches to one problem and developing of one's own thoughts occur in the context of Authority's realm out of a desire to confirm to Authority's demand. In Chap. 3, I will further discuss the difficulties in the measurement in terms of differentiating Position 4b from the next position, Position 5. The difficulties are in part due to the similarities in the behavioral patterns of individuals in these positions. Because of possible similarities between Positions 4b and 5, the boundaries between Positions 4 and 5 are not as clear in actuality.

Relativism
Position 5 Relativism Correlate, Competing, or Diffuse. In this position, all knowledge and values, including the ones of authorities (note here, Authorities become authorities), are perceived as relativistic. *Relativism* is the common characteristic whereas the right-wrong dualism becomes the special case in the new context. The revolution here is "the most violent accommodation of structure in the entire development" in that it involves a "complete transposition between part and whole, figure and ground..." (Perry 1970, p. 111). *"Relativism Correlate"* and *"Relativism Competing"* in the subtitle speak of some unresolved issues in the full transition, which is denoted as *"Relativism Diffuse."* Accompanying this revolution are new implications, such as the breakdown of old "guidelines and identity" and a new horizontal relation with authorities.

Position 6 Commitment[5] *Foreseen.* This position describes a realization of a necessity to commit oneself in a relativistic world. A relativistic world provides plenty of opportunities for one to exercise reason, but there is a limit in that "reason itself remains reflexively relativistic" (Perry 1970, p. 135). Acknowledging that reason itself cannot fully justify or assure her, she must commit herself through her own faith assuming at the same time the responsibility associated with the choice.

Commitment (within Relativism)
Position 7 Initial Commitment. A person in this position has undertaken his own responsibilities in some major area of his life, e.g., career and decision making about who he is or who he will be. Along with the decision, he also experiences a settled relief internally because of the initial commitment.

Position 8 Orientation in Implications of Commitment. A person in this position has begun to experience the implications of commitment. She also starts to explore the "subjective and stylistic issues" that are related with the implications of commitment; for example, while a person may find that decisions on who he is in a

---

[4]Relativism: "A plurality of points of view, interpretations... in which the structural properties of contexts and forms allow of various sorts of analysis, comparison and evaluation in Multiplicity" (Perry 1970, Glossary page). Compare Multiplicity.

[5]Commitment: "A conscious act", "An affirmation of personal values or choices in Relativism", "as distinct from commitment to an unquestioned, or unexamined belief, plan or value" (Perry 1970, Glossary page).

career choice actually turn out to be only a first step, there are many ongoing ways to approach this goal and the rest of other choices throughout life.

Position 9 Developing Commitment. A person in this position has developed a maturity in which his identity was affirmed in his commitment and the associated responsibilities. He is aware that commitment is an on-going activity "to a degree that is partly his own to determine and partly in the hands of fate, he is to be forever on the move" (Perry 1970, p. 154).

Perry's theory has been recognized as a piloting work in the understanding of epistemological knowing of college students (Hofer and Pintrich 1997; Moore 2002). This theory was originally developed based on a series of qualitative studies among Harvard college students in the 1950s and 1960s by Perry and his colleagues. The piloting studies that were performed in the 1950s with 31 students resulted in 98 longitudinal interviews with 17 complete four-year records. The enlarged studies in the 1960s included 109 students, resulting in 366 longitudinal interviews with 67 complete four-year reports. After this piloting study, several theories have been proposed. Each of these theories extended the original developmental theory by Perry from different perspectives.

The first half of the development described in Perry's theory is focused on intellectual development, and the second half is focused on ethical, moral, and identity development (Moore 2002). From the measurement and instrumental perspective, it can be difficult to separate these different stages. Most research done in instrumentation development focused on measuring the first five positions because the most dramatic change happens between *Multiplicity* and *Relativism* according to Perry's theory (Perry 1970). The difficulty in the measurement will be discussed in more details later in 2.5 Measurement of Epistemological Development.

2.3   Extension of Epistemological Development Frameworks After Perry's Theory

After the pioneering work by Perry on the intellectual and ethical development of college students, four other major models have emerged from empirical research in the literature body of personal epistemological development, as summarized by Hofer and Pintrich (1997) and Felder and Brent (2004). These include: (1) Belenky et al.'s work on "women's way of knowing" based on women from diverse backgrounds (1986), (2) Baxter Magolda's epistemological Reflection Model (1992), (3) King and Kitchener's Reflective Judgment model (1994), and (4) Kuhn's Argumentative Thinking (1991). The five models, including Perry's theory, each presents distinct points in the number and the naming of different positions/stages/perspectives. In the following section, each model will be reviewed in the order of time when a complete model was published. The discussions will include their main positions/stages/perspectives, their methods, and the implications and limitation of each model. Also later in the discussion, each model will be compared to the original Perry theory and other models in regards to the details of stages/positions/perspectives in each model. Based upon these review and discussions, and comparison across frameworks, a synthesis of different theoretical frameworks will be presented in the next section.

Women's Way of Knowing (Belenky et al. 1986)
One of the criticisms about Perry's work was that it was based on mainly white, elite college students. Belenky et al. were interested in women's knowing and learning. Using interview questions specifically related to gender, relationships, moral decisions, education, and knowledge, Belenky et al. interviewed 135 women—where 90 of these participants came from 6 diverse academic institutions and the additional 45 were participating in services provided by human service agencies on parenting. Using the metaphor of "Voice," Belenky et al. grouped women's ways of knowing into five categories:

1. *Silence.* A position in which women experience themselves as a voice-less, mind-less, and passive being and subject to external authority. They view themselves as "deaf and dumb" and as not being able to learn from others' words.
2. *Received Knowledge.* Women with this perspective perceive themselves as being able to listen to other's voices. They can receive knowledge from external authority and even reproduce knowledge. However, they believe that all knowledge originates from external authority; they themselves are not constructors of knowledge. Women who adhere to the perspective of received knowledge also believe an either-or dualism. There is no gray area to truth; it is either right or wrong.
3. *Subjective Knowledge.* Women with this perspective see truth as subjective, intuited, and personal. Truth is a private issue and cannot be forced on others. Belenky et al. (1986) noted that *Subjectivism* or (*Subjective Knowing*) is interchangeable with Perry's *Multiplicity* because of the emphasis on personal truth in both positions. The difference is that a woman with subjective knowing holds an anti-rationalist attitude and bases her knowing on intuition or personal experiences.
4. *Procedural Knowledge.* In this position, women start to realize that they cannot know everything by intuition or experiences. Knowing requires careful observation and reasoning. They are learning and applying procedures, techniques or skills of acquiring and communicating knowledge. Within *Procedural Knowledge*, two ways of knowing—*Separate Knowing* and *Connected Knowing* —were identified. The separate knower learns through excluding personal feelings and beliefs and adopting a different lens through explicit formal instruction, while the connected knower learns through being empathetic with the object (e.g., a poem)/person and adopting the lens of another person (in the case of a poem, the lens is that of the poet). Separate and connected knowers both learn to take off their own lenses and take up different ones. Although Hofer and Pintrich (1997) enlisted *Procedural Knowing* as parallel to Perry's *Relativism*, there are only a few indications showing that some parallel characteristics exist between *Procedural Knowing* and Perry's *Relativism*; for

example, in both positions, the knower starts to abandon the either-or thinking and takes into consideration alternative perspectives. However, *Procedural Knowing* cares only about the <u>form</u> of knowing, whereas Perry's *Relativism* is concerned to a greater extent with the <u>certainty</u> of knowing or knowledge.

5. *Constructed Knowledge*. Constructed knowing refers to an effort in which women try to reclaim the self by integrating intuitive knowledge and the knowledge that they learned from others. A constructed knower abandons either-or thinking completely and demonstrates a high tolerance of ambiguity and conflict. Their thought aligns with the basic thought of constructivism: "all knowledge is constructed, and the knower is an intimate part of the known" (p. 137).

The work by Belenky et al. extended the work of Perry by including women's ways of knowing, thereby providing a unique perspective for gender-related study. They also acknowledged that similar categories can be found in the original Perry model except the *Silence* position. However, researchers have noted a serious concern in Belenky et al.'s methodology in the study (Hofer and Pintrich 1997). On one hand, in the interview protocol by Belenky et al. a section on "Gender" and a section on "Relationships" were placed before sections on "Real Life Moral Dilemma," "Education," and "Ways of Knowing." On the other hand, their main findings focus on women's self-knowledge, the relation between self-knowledge and knowing, inner and outer voices, and the connected characteristics of women's way of knowing. Considering the methodology and findings, it is hard to tell the degree to which the interview questions have affected these main findings. Also, different from Perry's theory, which focuses more on the certainty or limits of knowledge, Belenky et al.'s women's way of knowing emphasizes more the source of knowledge, especially in relation to the self (Hofer and Pintrich 1997). Last but not least, Belenky et al. chose to study only women participants, which itself can be both an advantage to draw conclusions about women and a drawback open to criticism about the claims. Although they claim that some of the ways of knowing are not limited to women, e.g., connected knowing, no substantial evidence was provided to support these claims.

Epistemological Reflection (Baxter Magolda 1992)

Continuing on the same main line of epistemological development, Baxter Magolda based her research regarding epistemological development on Perry's model. Also, when combining the concerns raised by Belenky et al. (1986) about women's role and gender dimensions in epistemological development, she explored a gender-inclusive model of epistemological development for young adults from 18 to 30 years.

Using both interviews and the Measure of Epistemological Reflection (MER), which she developed via a series of studies (Baxter Magolda 1985, 1987, 1992; Baxter Mogolda and Porterfield 1988), she conducted a longitudinal study following 101 participants (51 men, 50 women) through their college years. She was able to continue following 70 out of the 101 participants in their post-collegiate phase. She was even following 39 of 70 participants by year 12. Out of these 39 participants, 16 were involved in advanced education (11 with a master's degree, 1 with a PhD degree, and 4 with other degrees, e.g., JD). The eight years' of research data regarding the post-collegiate participants extended Perry's original work to the post-collegiate population.

Using grounded theory, Baxter Magolda (1992) named the following four perspectives in her Epistemological Reflection Model: *Absolute Knowing, Transitional Knowing, Independent Knowing*, and *Contextual Knowing*. Gender-related patterns were identified within all but the last perspective. Gender-related patterns convey that each gender uses one pattern more than the other. However, these gender-related patterns are not exclusive to one gender.

Absolute Knowing: Receiving or Mastering Knowledge Learners who employ this perspective regard knowledge as absolute and certain. They learn the knowledge from authorities who know the truth. More women than men used the pattern of *Receiving Knowledge*. This pattern is parallel to *Received Knowledge* in Belenky et al.'s women's ways of knowing, in which learning takes place via listening and acquiring information. More men use the *Mastering Knowledge* pattern, which is similar to the experiences of male participants in Perry's study, where learners actively engaged in activities, debating, and quizzing of peers.

Transitional Knowing: Impersonal and Interpersonal Pattern Learners maintain the either-or thinking about knowledge in some disciplines, like mathematics and physics, as satisfactory, yet view knowledge in some other disciplines, like the humanities or social sciences, as uncertain. Within this perspective, these learners then focus on developing an understanding instead of acquiring knowing in these areas where knowledge is perceived as uncertain. *Impersonal and Interpersonal Patterns* were identified as gender-related patterns. More men use the *Impersonal Pattern*; more women use the *Interpersonal Patterns*.

Independent Knowing: Individual and Interindividual Pattern Learners with *Independent Knowing* believe that knowledge in itself is uncertain. Every individual thinker has his or her own viewpoints. Peers are encouraged to share views. Peers are also viewed as a source of knowledge along with authorities who are no longer viewed as the only resource for knowledge. Again, *Individual and Interindividual patterns* were identified from Baxter Magolda's findings to capture the gender-related patterns. Felder and Brent (2004) suggested that these two patterns are comparable to Belenky et al.'s *Separate Knowing* and *Connected Knowing* under the *Procedural knowledge* in women's ways of knowing.

Contextual Knowing: Learners with this perspective are able to judge the nature of knowledge based on evidence existent within different contexts. Learners both exercise thinking and compare different perspectives and ideas. Learners with this perspective believe that "some ideas are more valid than others" based on reasoning through available evidence (Baxter Magolda 1992, p. 170). Contextual knowers "think through problems," "integrating knowledge," and "apply it in a context" (Baxter Magolda 1992, p. 170). This perspective is very similar to the last positions of Perry's model (starting from Position 5 up to Position 9). It also parallels Belenky et al.'s *Constructed Knowledge*. No gender-related patterns were identified here because of the small number of participants. Baxter Magolda commented that it was possible that gender-related patterns converged at this position, in part because contextual knowing exhibits characteristics of both "connecting to others," which is the focus of the interindividual pattern, and "thinking independently," which is the focus of the individual approach.

Baxter Magolda's Epistemological Reflection model continued along the same main thread that was laid out by Perry. Built upon Perry and Belenky et al.'s former work, Baxter Magolda proposed a gender-inclusive model as a possible model of an epistemological development trajectory based upon a longitudinal study. Differing from Perry's male, elite college student population or Belenky's et al.'s female-only participants with a diverse educational background, Baxter Magolda focused on both the college-educated male and female participants. In this sense, Baxter Magolda extended the original Perry model and combined this with the gender perspective proposed by Belenky et al. and proposed a more gender-inclusive model.

However, it should be noted that participants of this study came from the same Midwestern university. The members of the population were mainly white (97 %) and mostly from middle-class families. Therefore, the extent to which this trajectory can be applicable to other races or ethnicities remains unclear.

Baxter Magolda extended the population into the body of post-collegiate young adults. Some of these individuals received advanced degrees, although only one of the participants obtained a PhD degree. This study may render some useful implication, yet still, the extent to which the epistemological developmental patterns can be similarly considered applicable to the doctoral-level students is unclear.

Finally, Perry's model focused on students from a liberal art college. Belenky et al.'s female participants came from a variety of educational levels and backgrounds. Baxter Magolda's participants were from a university with a liberal arts focus. None of these original studies focus on students with an engineering education background. The implications and applications of these epistemological models in engineering education will be discussed in Sect. 4.1.

Reflective Judgment Model (King and Kitchener 1994)
King and Kitchener focused on the epistemological assumptions and the reasoning processes of older adolescents and adults specifically when they face ill-structured problems. They also identified a trend in the development of reasoning skills that is similar, to a certain degree, with the above-mentioned cognitive developmental frameworks.

They conducted interviews centered on four ill-structured problems on topics such as the objectivity of news reporting. Also, they asked participants to justify their point of view in six follow-up questions. Based on their findings, they proposed their Reflective Judgment Model, which includes seven stages that are organized into three types of thinking: *Pre-reflective Thinking, Quasi-Reflective Thinking,* and *Reflective Thinking.* In each stage, they also defined the "view of knowledge" and "concept of justification" to best describe the characteristics of these stages and the specific type of thinking.

Pre-reflective Thinking includes three stages (Stages 1, 2, and 3). In Stage 1, knowledge is absolute and needs no justification. This type of thinking typically exists in young children but is not observed in King and Kitchener's studies. Stage 2 is similar to Perry's *Dualism* in which knowledge is assumed to be absolutely certain and is possessed by authority or to be temporarily unavailable. In Stage 3, beside the certainty of knowledge, there is a component of knowledge that is temporarily uncertain, and judgment itself is based on personal opinion.

Quasi-Reflective Thinking includes two stages (Stages 4 and 5). In Stage 4, knowledge is uncertain and knowing always involves ambiguity. Arguments and evidence that support this knowledge are idiosyncratic. Stage 5 features the subjectivity and context specificity of knowledge. "Other theories could be as true as my own, but based on different evidence" (King and Kitchener 2002, p. 42). Hofer and Pintrich (1997) stated that this stage resembles some characteristics of relativism by Perry (1970). However, King and Kitchener stated that individuals in Stage 5 "frequently appear to be giving a balanced picture of an issue or problem rather than offering a justification for their own beliefs," i.e., that "individuals are able to relate and compare evidence and arguments in several contexts" while however, they still cannot "coordinate evidence and arguments across context into a simple system." This is different from Perry's *Relativism*, *Constructed knowledge* (Belenky 1986), or *Contextual knowing* (Baxter Magolda 1992).

Reflective Thinking (Stages 6 and 7) parallels *Relativism* (Perry 1970), *Constructed knowledge* (Belenky 1986), and *Contextual knowing* (Baxter Magolda 1992) because, in this state, knowledge is an outcome of a process in which different solutions or evidence and perspectives are evaluated.

In King and Kitchener's 20 years of longitudinal study, they studied participants ranging in age from their teenage years to middle adulthood. Their studies with early level graduate students showed mean scores between Stage 4 and Stage 5. Their studies with an advanced level of graduate students showed mean scores between Stage 5 and Stage 6. Stage 6 reasoning has only been typically observed among advanced doctoral students.

Again, King and Kitchener's framework focused more on the application of knowledge beliefs in making justifications in the process of solving ill-structured problems. Their study focus is more on the thinking and justification process instead of on the applicants' beliefs about knowledge. Therefore, it may not be directly comparable to the first three models although it did suggest a similar epistemological developmental trend.

Argumentative Reasoning (Kuhn 1991)
The primary purpose of Kuhn's study was to understand argumentative thinking, but the process of trying to understand argumentative thinking or reasoning has also offered insights into epistemological perspectives. In Kuhn's study, she included a broad sample of participants ranging in age group from the teens, 20s, 40s, and up to 60s with 40 subjects in each division. Participants were interviewed regarding their reasoning process concerning three ill-structured, real-life social problems such as unemployment. Several sections of her interviews asked questions that were

related to epistemological perspectives about expertise, multiple viewpoints, and certainty of knowledge, such as, "Do experts know for sure what causes...?"

Kuhn reported epistemological thoughts observed among participants along the same line laid out by Perry (1970), Belenky et al. (1986), Baxter Magolda (1992), and King and Kitchener (1994). She presented her theory as three categories of epistemological views: *Absolutist*, *Multiplist*, and *Evaluatist*.

The Absolutist viewpoint resembles that of *Dualism* taken by Perry, in which knowledge is certain and absolute. The *Multiplist* viewpoint features a radical subjectivity, which resembles that of *Subjective Knowledge* proposed by Belenky (1986), in which more weight is given to subjective knowledge and emotions rather than to facts. In essence, a *multiplist* believes that all individuals' views have equal legitimacy as that of an expert. An individual taking the *evaluative* viewpoint believes that different views need to be compared and evaluated concerning their merits.

These three positions have again mapped the main line of the epistemological developmental trend from a dualistic view to a more sophisticated way of evaluating evidence and views. This pattern, which has been repeatedly validated by multiple researchers throughout the years (Perry 1970; Belenky et al. 1986; Baxter Magolda 1992; King and Kitchener 1994; Kuhn 1991), will be summarized in the next section and used as the theoretical framework for this work.

2.4 Synthesis of Theoretical Frameworks

The major themes and findings of each model demonstrate that they are each focused on different landscapes of epistemological development. For example, Perry and his colleagues (1970) initially derived their theory from a predominantly male model, whereas, Belenky et al. (1986) focused on an exclusively female sample and described women's way of knowing using the metaphor of "voice." As another example, the first three models centered on mapping a more or less developmental and structural sequence and the last two models focused on the influences of epistemological assumptions on the thinking processes. Despite these distinct points, however, the parallel positions/stages/perspectives were evident across five models in that they all suggested *a movement from a dualistic view of knowledge to a contextual, constructivist perspective*. This movement, which was originally depicted in Perry's model, remains the main thread across the four additional models. They each presented significant extensions to the intellectual development of "the same fundamental journey" (Moore 2002, p. 23).

Therefore, I shall keep primarily the naming from Perry's theory for the theoretical framework for this work. An ongoing refinement of Perry's model grouped Perry's nine positions into four major categories: *Dualism* (Positions 1 and 2), *Multiplicity* (Positions 3 and 4), *Relativism* (Positions 5 and 6), and *Commitment (within Relativism)* (Positions 7 through 9) (Culver and Hackos 1982; Knefelkamp 1974; Knefelkamp and Slepitza 1978; Moore 1991, 1994, 2002). Each category in

**A Similar Trend of Epistemological Development
Suggested among the Five Theoretical Frameworks**

Epistemological Development	Dualism	Multiplicity				Relativism	Commitment
Intellectual and Ethical Development — Perry, 1970	Dualism	Multiplicity Subordinate		Multiplicity Correlate		Relativism	Commitment
Women's Way of Knowing — Belenky et al., 1986	Received Knowledge	Subjective Knowledge		Separate Knowing / Connected Knowing — Procedural Knowledge		Constructed Knowledge	
Epistemological Reflection — Baxter Magolda, 1992	Mastering Knowledge / Receiving Knowledge — Absolute Knowing	Impersonal Pattern / Interpersonal Pattern — Transitional Knowing		Individual Pattern / Interindividual Pattern — Independent Knowing		Contextual Knowing	
Reflective Judgment — King and Kitchener, 1994	Pre-Reflective Thinking	Quasi-Reflective Thinking				Reflective Thinking	
Argumentative Reasoning — Kuhn, 1991	Absolutionist	Multiplist				Evaluatist	

Theoretical Frameworks (vertical axis label)

Fig. 2.2 Overview of five current epistemological development frameworks for late adolescents and adults (Zhu and Cox 2015; modified from Hofer and Pintrich 1997, p. 92 and Felder and Brent 2004, p. 5)

this theory corresponds to different stages/positions/perspectives of other cognitive developmental theories, including Belenky et al.'s Women's ways of thinking (1986), Baxter Magolda's Epistemological reflection (1992), King and Kitchener's Reflective Judgment Model (1994), and Kuhn's Argumentative Reasoning (1991).

An alignment of the five models was modified and refined from former literature review works (Hofer and Pintrich 1997; Felder and Brent 2004). Hofer and Pintrich (1997) provided an overview as to the alignment of stages/positions/perspectives across the five models. Felder and Brent (2004) also attempted to establish such an alignment in greater detail by taking into consideration the gender-related patterns. Based on prior works, a diagram (Fig. 2.2) was created to demonstrate the alignment of the five theoretical frameworks to provide an overview of the aligning across these different theories (Zhu and Cox 2015). It should be noted that the representations and comparisons of different stages/positions/perspectives across five frameworks are just illustrations to facilitate understanding. In actuality, there are various overlaps or intertwining among stages/positions/perspectives. They are not to be viewed as precisely confined, separated stages/positions/perspectives as they may appear in the diagram.

The theoretical framework used in this research is refined based on these models using mainly the naming from the original Perry's theory and also definitions of other models in regards to the details of stages/positions/perspectives in each model. As mentioned above and illustrated in Fig. 2.2, the ongoing refinement of Perry's model suggests the grouping of Perry's nine positions into four major categories: *Dualism* (Positions 1 and 2), *Multiplicity* (Positions 3 and 4), *Relativism* (Positions 5 and 6), and *Commitment (within Relativism)* (Positions 7 through 9) (Culver and Hackos 1982; Knefelkamp 1974; Knefelkamp and Slepitza 1978; Moore 1991, 1994, 2002). Considering the similarities and potential gender-related patterns, the process of epistemological development is laid out as follows. It should be noted that gender-related patterns will be taken into consideration in the theoretical framework and data analysis in the later stage of the research, although they are not specified in the framework or in the diagram.

Across the five current models, only Belenky et al.'s (1986) Women's ways of knowing observed the "silence" stage found among underprivileged women. Considering the Chinese engineering doctoral students, the participants in this work, this particular position may not be as relevant for these students who did possess a certain level of academic achievement when compared to the under-privileged women in Belenky et al.'s study. Therefore, this stage is omitted in the theoretical framework used in this research.

Dualism

In the first stage of epistemological development, all five of the cognitive developmental theories found a dualistic way of knowing or thinking among their participants. Belenky et al. (1986) noted that their *Received knowledge* is similar to Perry's *Dualism*. However, in Perry's *Dualism*, a person identifies "We" with "Authority-right" and "They" with "Illegitimate-wrong;" for women with a *Received Knowledge* perspective, they do not align themselves as close to the authorities. Also, "We" and "They" are intertwined. *Received Knowledge* in Belenky et al.'s Women's ways of knowing is paralleled to *Absolute Knowing: Receiving Knowledge* in Epistemological reflection (Baxter Magolda 1992), in which learning takes place via listening and acquiring information. On the other hand, more men used the mastering pattern, which is similar to the experiences of male participants in Perry's study, where learners actively engaged in activities, and the debating and quizzing of peers. Therefore, Perry's *Dualism* is viewed as parallel to *Absolute Knowing: Mastering Knowledge* (Felder and Brent 2004). To summarize, this stage of thinking parallels Baxter Magolda's *Absolute Knowing*. Despite the gender-related patterns, all of these frameworks emphasized the dualistic thinking of a person at this stage. Therefore, the first stage is named *Dualism*.

Multiplicity

The position of *Multiplicity* by Perry (1970) is listed in Fig. 2.2 as *Multiplicity Subordinate* and *Multiplicity Correlate* because the characteristics of these two positions are comparable to stages in other models (Felder and Brent 2004). Perry's *Multiplicity Subordinate* is comparable to the *Impersonal Pattern* of *Transitional Knowing* (Baxter Magolda 1992) and *Subjective Knowledge* (Belenky et al. 1986) is comparable to the *Interpersonal Patterns* of *Transitional Knowing* (Baxter Magolda 1992) (refers to Felder and Brent 2004, p. 5, Table 1). The view of Felder and Brent (2004) takes into account the gender-related pattern and presents a reasonable comparison across the first three models.

Felder and Brent then paralleled Perry's *Multiplicity* to the male-related patterns within *Procedural Knowledge* (Belenky et al. 1986) and *Independent Knowing* (Baxter Magolda 1992). Hofer and Pintrich (1997) presented a different idea by paralleling *Independent Knowing* (Baxter Magolda 1992) to Perry's next stage—*Relativism*. Although *Independent Knowing* indeed does have some overlaps with Perry's *Relativism* in that they both suggest the legitimacy of students, and not just the authorities, may be viewed as a source of knowledge, nevertheless, *Independent Thinking* is still considered as parallel to *Multiplicity*. According to a later reflection by Baxter Magolda about her own model (Baxter Magolda 2002, p. 100), she suggested that the core of *Relativism* or relativistic thinking was much closer in conceptualization to *Contextual Knowing* than *Independent Thinking*.

It should be noted here that Perry's *Relativism Subordinate*, which was originally defined as part of *Multiplicity*, is not included here because of some complexities in measurement that will be further discussed in Chap. 3. *Relativism Subordinate* is not listed in the diagram although it should be part of *Multiplicity* conceptually.

To summarize, I use the term *Multiplicity* as the naming of this stage. It covers *Multiplicity Subordinate* and *Multiplicity Correlate* defined by Perry (1970), *Subjective* and *Procedural Knowledge* by Belenky et al. (1986), and *Transitional* and *Independent Thinking* by Baxter Magolda (1992) (as shown in Fig. 2.2). It is comparable to King and Kitchner's (1994) *Quasi-Reflective Thinking* and Kuhn's (1991) *Multiplist* in that it emphasizes an equal absolutism the understanding of knowledge—"Everyone has a right to his own opinion" (Perry 1970, p. 97).

Contextual Relativism

After *Multiplicity,* there shall follow the most important transition to the critical stage of the relativistic type of thinking, one that resembles *Constructed Knowledge* (Belenky et al. 1986), *Contextual Knowing* (Baxter Magolda 1992), *Reflective Thinking* (King and Kitchener 1994), or *Evaluatist* (Kuhn 1991). To capture the essence of this stage of thinking, the term "*Contextual Relativism*" is used here to emphasize that a person in this position views knowledge as "contextual and relativistic" (Perry 1970, p. 109), i.e., the concept that knowledge claims become the outcome of a constructed process by comparing and weighing different contexts and evidence. Justifications come from comparing different contexts, evidence, solutions, possible consequences, and so on. This stage will still be called *Relativism* in the other sections of this document.

Commitment (within Contextual Relativism)

Last but not least, *Commitment (within Contextual Relativism)* may be seen as parallel to *Commitment (within Relativism)* by Perry. This stage is not clearly defined in other models. In Perry's original model, it is also the one least understood or elaborated upon by Perry (Baxter Magloda 1985). *Commitment (within Contextual Relativism)* is so far the least explored stage in the literature, partly because of the lack of intense research in more advanced adult populations with higher degrees, education, or life experiences, for example, the doctoral student population (Moore 2002). This stage will still be called *Commitment within Relativism* in the other sections of this document.

As a final note to this synthesis of theoretical framework section, besides the above-mentioned theoretical frameworks, an additional framework was hypothesized by Schommer (1990, 1993), named *Epistemological Beliefs. Epistemological Beliefs* include a set of beliefs about knowledge and knowing, specifically, structure, certainty, source of knowledge, and control and speed of knowledge acquisition. Different from the epistemological developmental frameworks described above, the five dimensions are not organized in an overall developmental order. Therefore, this view was not listed in Fig. 2.2. The first three dimensions originated from Perry's work; the last two originated from Dweck and Leggett's nature of intelligence (1988) and Schoenfeld's beliefs about mathematics (1983, 1985, 1988). This view has encountered methodological and theoretical difficulties. The methodological difficulties will be discussed in the next section. For the theoretical concerns, Hofer and Pintrich (1997) and others (Debacker et al. 2008) have raised questions about the validity of this theoretical hypothesis. Although some dimensions, e.g., the structure and certainty of knowledge, appear to be epistemic, other dimensions, e.g., control and speed of knowledge acquisition, fall outside the conceptualization of epistemological beliefs (see 2.1 Conceptualization of Personal Epistemology). Factor analysis in their studies has also shown that some dimensions appear not to be following the patterns of other dimensions (Schommer 1990). This may be interpreted as "evidence that the dimensions operate independently," but it may be an indication of lacking relations between these two factors with other dimensions (Hofer and Pintrich 1997, p. 108). Given these unsettled issues in *Epistemological* Beliefs, it is not considered in this work as a theoretical framework.

To summarize, Perry's theory and its extensions over the past four decades have repeatedly revealed a similar epistemological developmental trend despite some distinct points and emphases in each model. They all have suggested "movements from a dualistic, objectivist view of knowledge to a more subjective, relativistic stance and ultimately to a contextual, constructivist perspective of knowing" (Hofer 2002, p. 7). Also, there is no compelling evidence that these frameworks present distinct theories (Moore 2002). As Moore commented in his review about the impact of Perry's model 30 plus years after its first publication, "Even after thirty years of extensive and varied scholarship, the Perry's theory continues to reflect the most critical dimension to educators' understanding of learning and students' approaches to learning" (Moore 2002, p. 18).

References

Baxter Magolda, M. B. (1985). A new approach to assessing intellectual development on the perry scheme. *Journal of College Student Personnel, 26*, 343–351.

Baxter Magolda, M. B. (1987). Comparing open-ended interviews and standardized measures of intellectual development. *Journal of College Student Personnel, 28*, 443–448.

Baxter Magolda, M. B. (1992). *Knowing and reasoning in college.* San Francisco: Jossey-Bass.

Baxter Magolda, M. B., & Porterfield, W. D. (1988). *Assessing intellectual development: The link between theory and practice.* Alexandria, VA: American College Personnel Association.

Baxter Magolda, M. B. (2002). The reflective judgment model: Twenty years of research on epistemic cognition. In B. K. Hofer & P. R. Pintrich (Eds.), *Personal epistemology: The psychology of beliefs about knowledge and knowing.* NJ. Erlbaum, Mahwah.

Belenky, M. F., Clinchy, B. M., Goldberger, N. R., & Tarule, J. M. (1986). *Women's ways of knowing: the development of self, voice and mind.* New York: Basic Books.

Culver, R. S., & Hackos, J. T. (1982). Perry's model of intellectual development. *Engineering Education, 72,* 221–226.

DeBacker, T. K., Crowson, H. M., Beesley, A. D., Thoma, S. J., & Hestevold, N. L. (2008). The challenge of measuring epistemic beliefs: An analysis of three self-report instruments. *The Journal of Experimental Education, 76,* 281–312.

Dweck, C. S., & Leggett, E. L. (1988). A social-cognitive approach to motivation and personality. *Psychological Review, 95*(2), 256–273.

Felder, R. M., & Brent, R. (2004). The intellectual development of science and engineering students. 1. Models and challenges, *Journal of Engineering Education, 93*(4), 269–277.

Flavell, J. H. (1979). Metacognition and cognitive monitoring. *American Psychologist, 34,* 906–911.

Hofer, B. K. (2001). Personal epistemology research: Implications for learning and teaching. *Journal of Educational Psychology Review, 13,* 353–383.

Hofer, B. (2002). Personal epistemology as a psychological and educational construct: An introduction. In B. Hofer & P. Pintrich (Eds.), *Personal epistemology: The psychology of beliefs about knowledge and knowing.* Mahwah, NJ: Lawrence Erlbaum Associates.

Hofer, B. K., & Pintrich, P. R. (1997). The development of epistemological theories: Beliefs about knowledge and knowing and their relation to learning. *Review of Educational Research, 67*(1), 88–140.

King, P. M., & Kitchener, K. S. (1994). *Developing reflective judgment: Understanding and promoting intellectual growth and critical thinking in adolescents and adults.* San Francisco: Jossey-Bass.

King, P. M., & Kitchener, K. S. (2002). The reflective judgment model: Twenty years of research on epistemic cognition. In B. K. Hofer & P. R. Pintrich (Eds.), *Personal epistemology: The psychology of beliefs about knowledge and knowing.* NJ: Erlbaum, Mahwah.

Kitchener, K. S. (1983). Cognition, metacognition, and epistemic cognition. *Human Development, 26,* 222–232.

Knefelkamp, L. L. (1974). Developmental instruction: Fostering intellectual and personal growth of college students. *Doctoral Dissertation,* University of Minnesota, Minneapolis. (Dissertation Abstracts 36, 3: 1271A. 1975).

Knefelkamp, L. L., & Slepitza, R. L. (1978). A cognitive-developmental model of career development: An adaptation of the Perry scheme. In C. A. Parker (Ed.), *Encouraging development in college students* (pp. 135–150). Minneapolis: University of Minnesota Press.

Kuhn, D. (1991). *The skills of argument.* Cambridge, England: Cambridge University Press.

Kuhn, D. (1999). A developmental model of critical thinking. *Educational Researcher, 28*(2), 16–26.

Kuhn, D. (2000). Metacognitive development. *Current Directions in Psychological Science, 9,* 178–181.

Moore, W. S. (1991). *The Perry scheme of intellectual and ethical development: An introduction to the model and two major assessment approaches.* Paper prepared for the annual meeting of the American Educational Research Association, Chicago, IL.

Moore, W. S. (1994). Student and faculty epistemology in the college classroom: The Perry scheme of intellectual and ethical development. In K. Pritchard & R. M. Sawyer (Eds.), *Handbook of college teaching* (pp. 45–67). Westport, CT: Greenwood Press.

Moore, W. S. (2002). Understanding learning in a postmodern world: Reconsidering the Perry scheme of intellectual and ethical development. In B. Hofer & P. Pintrich (Eds.), *Personal epistemology: The psychology of beliefs about knowledge and knowing.* Mahwah, NJ: Lawrence Erlbaum Associates.

Perry, W. G. (1970). *Forms of intellectual and ethical development in the college years: A scheme.* New York: Holt, Rinehart and Winston.

Piaget, J. (1950). *Introduction a l'epistemologie genetique.* Paris: Presses Univ. de France.

Piaget, J. (1972). English Translation of *The principles of genetic epistemology* (Translated from the French by W. Mays). London: Routledge.

Schoenfeld, A. (1983). Beyond the purely cognitive: Belief systems, social cognitions, and metacognitions as driving forces in intellectual performance. *Cognitive Science, 7*(4), 329–363.

Schoenfeld, A. H. (1985). *Mathematical problem solving.* San Diego, CA: Academic Press.

Schoenfeld, A. (1988). When good teaching leads to bad results: The disasters of "well taught" mathematics classes. *Educational Psychologist, 23*(2), 145–166.

Schommer, M. (1990). Effects of beliefs about the nature of knowledge on comprehension. *Journal of Educational Psychology, 82*(3), 498–504.

Schommer, M. (1993). Epistemological development and academic performance among secondary students. *Journal of Educational Psychology, 85,* 406–411.

Zhu, J., & Cox, M. F. (2015). Epistemological development profiles among chinese engineering doctoral students in U.S. institutions: An application of Perry's Theory. *Journal of Engineering Education, 104,* 243–362.

Chapter 3
Measurement of Epistemological Development

This chapter provides an overview of multiple methods and tools that have been used in measuring epistemological development, including both qualitative and quantitative methods.

3.1 Qualitative Measures of Epistemological Development

Throughout the past 40 years of research in the cognitive developmental field, primarily, qualitative methods are mostly acknowledged and recommended for providing in-depth, rich details of students' epistemological understanding. It has been adopted as a more reliable way by which to understand the students' cognitive developmental stage as portrayed in various former studies, including those of Perry (1970), Belenky et al. (1986), Baxter Magolda (1992). The drawback of using qualitative methods in these prior studies obviously is the intensiveness of one-on-one interviews and the data analysis process. In addition to the intensiveness of data collection and analysis, the nature of qualitative research limits the possibility to perform a large-scale estimation of the cognitive development of a large student population. For each series of studies, I will summarize a brief discussion about their interview process and methods.

The central concern of Perry's work was students' own reports of their college experiences as expressed in the students' own languages. Therefore, phenomenology was used to guide their interview process. To avoid biasing the students' thought structure through the structure of the interview protocol, interviews were conducted in as open-ended a way as possible. The interview was started with the simple question: "What stands out for you about the year?" (Perry 1970, p. 19) and refused to provide further specific questions. This attempt aimed at encouraging the participants to express varied ideas and to make sense of their personal experiences in their own way. This attempt and the genuine interest in uncovering the students' own experiences through their own language were compensated by the

© Springer Science+Business Media Singapore and Higher Education Press 2017   29
J. Zhu, *Understanding Chinese Engineering Doctoral Students in U.S. Institutions*,
East-West Crosscurrents in Higher Education, DOI 10.1007/978-981-10-1136-8_3

students' taking the lead in expressing their ideas. Perry acknowledged that this procedure caused the beginning moments to be "socially awkward." However, as Perry stated, this procedure was crucial to the research in that the primary focus of the research was the participants' own way of making sense of his or her experience. As a result, this method also proved to be effective in helping the many students to express their personal ideas.

For Belenky et al.'s research (1986), they organized their interview protocol under different sections. As I noted under the discussion about their theoretical framework, there was potentially a serious concern in their methodology (Hofer and Pintrich 1997). In the interview protocol, a section on "Gender" and a section on "Relationships" were placed before sections on "Real Life Moral Dilemma," "Education," and "Ways of Knowing." Considering their main findings focused on women's self-knowledge, the relationship between self-knowledge and knowing, inner and outer voices, and the connected characteristics of women's way of knowing, it is hard to tell the degree to which the interview questions have affected these main findings.

For Baxter Magolda's studies (1992), she retained the interview method. In addition, she classified interview questions into six content areas based on previous work by Perry (1970), Kurfiss (1977). These six content areas included (1) the role of learners, (2) role of instructors, (3) role of peers, (4) perception of the evaluation of their work, (5) nature of knowledge, and (6) educational decision-making. Within each content area, she retained an un-structured interview procedure, i.e., where interview questions were asked to introduce the topic but not frame the response. By segregating different content areas, she was then able to present a clearer picture of students' views in each of the researched areas. Considering the advantage of this interview structure, this interview structure will be adopted as part of the methodological design of this work (see Chap. 5).

Reflective Judgment Interviews (RJIs) were developed by King and Kitchener (1994). The Reflective Judgment Interviews have five standard problems on questions like the objectivity of news or food safety. They also have additional discipline-based questions (Psychology, Chemistry, and Business). Based on the interviewee's understanding and justifications of the claims, raters will rate the interviewee as responsive with regards to two aspects: (1) the nature of knowledge and (2) the nature of justification. As discussed under the theoretical framework section, it has been widely used by a large number of participants (King and Kitchener 2002). However, this process may lead to concerns with regards to disciplinary issues. An intensive study in one discipline may potentially affect their reflective thinking in that discipline. Therefore, this measure may produce artifacts on measures across disciplines.

3.2 Quantitative Measures of Epistemological Development

To obtain data in large student populations and enable large-scale long-term experiments and comparisons, there have been several different pen-and-paper based measurement methods since the first issue of Perry's intellectual and ethical model. These pen-and-paper based measurement methods include: Baxter Magolda's Measure of Epistemological Reflection (MER) (1985, 1987, 1992), Learning Environment Preferences (LEP) and Measure of Intellectual Development (MID) (Moore 1989), King and Kitchener's Reasoning about Current Issues Test (RCI) (1994), Scale of Cognitive Devleopment (Fago 1995), Schommer's Survey of Epistemological Beliefs (1990), and Zhang's Cognitive Development Inventory (ZCDI, 1995, 1997) (note: the latest version, the fourth version of the ZCDI was provided to me by Dr. Lifang Zhang. ZCDI will be discussed separately in 4.3 Application of Epistemological Developmental Theory in Chinese Students section). These different measures have enabled large-scale studies and provided some valuable empirical data about cognitive development across gender, ethnicity, and also other variables such as personality, critical thinking skills, learning styles, and so on. However, most of these scales still require well-trained raters to effectively rank the results because these measures apply narrative-test responses to the pre-defined problems. This design largely limits the possible usage of most of these tests among larger population. Moreover, other limitations exist among some of these scales. For example, the RCI test has consistently provided scores that are more conservative than interview measures (King and Kitchener 2002). In MER, Baxter Magolda used only Positions 1–5 from Perry's model, in that Position 5 was deemed to be a logical transition and milestone toward relativism (1987). Therefore, she only validated MER in terms of Perry's Position 1 to Position 5. However, no additional information is available to test the rest of Perry's positions (Baxter Magolda 1987). The Measure of Intellectual Development was shown to give conservative scores, possibly even one or two Perry positions lower (Pavelich and Fitch 1988).

As I noted at the end of last section, Schommer (1990, 1993) hypothesized a framework named Epistemological Beliefs. This framework uses a more quantitative view than all of the five above-mentioned models. Schommer developed a 63-item questionnaire (Epistemological Questionnaire, EQ) to measure the five hypothesized dimensions. The development of this method has allowed researchers to perform large-scale measurements because so far there are very few survey instruments available to measure epistemological beliefs. Several researchers (Hofer and Pintrich 1997; DeBacker et al. 2008) have expressed a number of methodological concerns with this instrument. First, their factor analysis was performed using 12 subsets of items as variables organized by three educational psychologists prior to piloting (and not the original 63 items) (Schommer 1990). Second, the factor analysis generated four factors (Fixed Ability, Quick Learning,

Simple Knowledge, and Certain Knowledge), which were different from the original hypothesis.

In the literature, there are two other major instruments that were modified from the EQ and have gained attention among researchers: (1) Epistemological Beliefs Inventory (EBI) (Schraw et al. 2002) and (2) Epistemological Beliefs Survey (EBS) (Wood and Kardash 2002).

Researchers have tried to confirm the original framework by Schommer (1990, 1993) to a greater extent by organizing the items according to the original five structures. A 28-item EBI was constructed according to the definition of each epistemic dimension described by Schommer (1990) with seven items adapted from EQ via several pilot studies, content analysis, and revisions (Bendixen et al. 1998; Schraw et al. 2002). Their factor analysis among 160 undergraduates resulted in five factors, labeled as: Simple Knowledge, Certain Knowledge, Quick Learning, Fixed Ability, and Omniscient Authority. The internal consistency ranged from 0.58 to 0.68. Another later study, however, was not able to produce all five factors (Nussbaum and Bendixen 2003). In addition, the sample sizes in these tests were modest; n was usually less than 200. Later, Debacker et al. (2008) used two samples ($n1 = 378$ and $n2 = 417$) to test the psychometric properties of both EBI and EBS. They found a slightly better internal consistency and factor loading ratio for EBS than EBI.

Developers of EBS retained Schommer's items (1990) and tried to find a more stable factor structure among them in response to the concern raised by Hofer and Pintrich (1997). They combined all of Schommer's 63 items and a related measure by Jehng et al. (1993) and ran a factor extraction of all items. They also tried to examine whether these items would lead to the factors proposed by Schommer and to determine how the emergent factors would correlate with each other. The results among 793 participants lead to a five-factor solution. These five factors were labeled as: Speed of Knowledge Acquisition, Structure of Knowledge, Knowledge Construction and Modification, Characteristics of Successful Students, and Attainability of Objective Truth. Some of these factors' descriptions were similar to Schommer's original factors. For example, the Speed of Knowledge Acquisition overlaps with Schommer's Quick Learning factor. However, some factors seem novel from the original factor list. For example, the Structure of Knowledge and Knowledge Construction and Modification were novel factors that were not clearly identified in the original test run through EQ.

A closer examination of the definition of these five factors has shown that higher scores in the Knowledge Construction and Modification factor relate closely to the participants' epistemic development from a dualistic view to a more constructivist view. Here is a direct excerpt from Wood and Kardash (2002)'s descriptions of their emergent factors:

*(Factor 3) "Knowledge Construction and Modification" reflected participants' awareness that knowledge can be acquired and modified through strategies such as integrating information from various sources, reorganizing information according to a personal scheme, questioning information, and recognizing the tentativeness of information. High scores on this factor reflect the ideas that knowledge is constantly evolving, **is actively and***

personally constructed, and should be subjected to questioning. By contrast, low scores on this factor reflect a view that knowledge is certain, passively received, and accepted at face value. (p. 250)

For their descriptions, this factor appears reflect the epistemological trend repeatedly observed by Perry (1970), Belenky et al. (1986), Baxter Magolda (1992), King and Kitchener (1994), Kuhn (1991). Although EBS was not developed within Perry's framework, scores derived under this factor do serve as a useful indication for students' epistemological development in their knowledge construction and modification. The internal reliability for this subscale was reported in Debacker et al. (2008)'s study as 0.67 (Sample 1, $n = 380$) and 0.65 (Sample 2, $n = 415$). Therefore, this subscale will also be used for this work.

References

Baxter Magolda, M. B. (1985). A new approach to assessing intellectual development on the Perry scheme. *Journal of College Student Personnel, 26*, 343–351.

Baxter Magolda, M. B. (1987). Comparing open-ended interviews and standardized measures of intellectual development. *Journal of College Student Personnel, 28*, 443–448.

Baxter Magolda, M. B. (1992). *Knowing and reasoning in college*. San Francisco: Jossey-Bass.

Belenky, M. F., Clinchy, B. M., Goldberger, N. R., & Tarule, J. M. (1986). *Women's ways of knowing: the development of self, voice and mind*. New York: Basic Books.

Bendixen, L. D., Schraw, G., & Dunkle, M. E. (1998). Epistemic beliefs and moral reasoning. *The Journal of Psychology, 132*, 187–200.

DeBacker, T. K., Crowson, H. M., Beesley, A. D., Thoma, S. J., & Hestevold, N. L. (2008). The challenge of measuring epistemic beliefs: An analysis of three self-report instruments. *The Journal of Experimental Education, 76*, 281–312.

Fago, G. C. (1995). A scale of cognitive development: Validating Perry's scheme.

Hofer, B. K., & Pintrich, P. R. (1997). The development of epistemological theories: Beliefs about knowledge and knowing and their relation to learning. *Review of Educational Research, 67*(1), 88–140.

Jehng, J.-C. J., Johnson, S. D., & Anderson, R. C. (1993). Schooling and students' epistemological beliefs about learning. *Contemporary Educational Psychology, 18*, 23–35.

King, P. M., & Kitchener, K. S. (1994). *Developing reflective judgment: Understanding and promoting intellectual growth and critical thinking in adolescents and adults*. San Francisco: Jossey-Bass.

King, P. M., & Kitchener, K. S. (2002). The reflective judgment model: Twenty years of research on epistemic cognition. In B. K. Hofer & P. R. Pintrich (Eds.), *Personal epistemology: The Psychology of beliefs about knowledge and knowing*. NJ: Erlbaum, Mahwah.

Kuhn, D. (1991). *The skills of argument*. Cambridge, England: Cambridge University Press.

Kurfiss, J. (1977). Sequentiality and structure in a cognitive model of college student development. *Developmental Psychology, 13*, 565–571.

Moore, W. S. (1989). The learning environment preferences: Exploring the construct validity of an objective measure of the Perry scheme of intellectual development. *Journal of College Student Development, 30*, 504–514.

Nussbaum, E. M., & Bendixen, L. D. (2003). Approaching and avoiding arguments: The role of epistemological beliefs, need for cognition, and extraverted personality traits. *Contemporary Educational Psychology, 28*, 573–595.

Pavelich, M. J., & Fitch, P. (1988). Measuring students' development using the Perry model. In *1988 Proceedings of the American Society for Engineering Education*. Washington, DC.

Perry, W. G. (1970). *Forms of intellectual and ethical development in the college years: A scheme*. New York: Holt, Rinehart and Winston.

Schommer, M. (1990). Effects of beliefs about the nature of knowledge on comprehension. *Journal of Educational Psychology, 82*(3), 498–504.

Schommer, M. (1993). Epistemological development and academic performance among secondary students. *Journal of Educational Psychology, 85*, 406–411.

Schraw, G., Bendixen, L. D., & Dunkle, M. E. (2002). Development and validation of the epistemic belief inventory. In B. K. Hofer & P. R. Pintrich (Eds.), *Personal epistemology: The psychology of beliefs about knowledge and knowing* (pp. 103–118). Mahwah, NJ: Erlbaum.

Wood, P., & Kardash, C. (2002). Critical elements in the design and analysis of studies of epistemology. In B. K. Hofer & P. R. Pintrich (Eds.), *Personal epistemology: The psychology of beliefs about knowledge and knowing* (pp. 231–260). Mahwah, NJ: Erlbaum.

Zhang, L. F. (1995). The construction of a Chinese language cognitive development inventory and its use in a cross-cultural study of the Perry Scheme. Retrieved from ProQuest, The University of Iowa.

Zhang, L. F. (1997). *The Zhang cognitive development inventory* (Unpublished text). The University of Hong Kong, Hong Kong.

Chapter 4
Personal Epistemology in Application

Built upon the prior discussions on the theories and tools for personal epistemology, this chapter explores some piloting work in applying epistemological developmental theories among students.

4.1 Application of Epistemological Developmental Theories in Engineering Students

In 1982, Culver and Hackos published an article about applying Perry's model in engineering education. They suggested that the curriculum and instructional methods then in use, which had a concentration on content-based courses in engineering undergraduate education, had not effectively helped the students to progress in intellectual development. Instead, because of the authority-oriented instructional methods, the lack of integration across courses, and the lack of problem-solving opportunities that were seen within the many different courses, engineering students were possibly delayed in their intellectual development. After his work, there have been quite a few researchers who have either used Perry's model, tried to understand engineering undergraduate students' intellectual development, or developed innovative instructional measures to facilitate students' development along Perry's scale in engineering education (Culver et al. 1990; Pavelich and Moore 1993, 1996; Wise et al. 2004).

Pavelich and Moore (1996) found that by graduation only one quarter of engineering students reached Position 5 in Perry's nine-position scale; one third of these students fell below Position 4. The overall average for seniors was 4.28 ± 0.70. Wise et al. (2004) also obtained similar findings for seniors with an average of 4.2 ± 0.50. Wise et al. further pointed out a faster development of nearly one Perry's position from juniors to seniors. The researchers suggested that the possible cause could be that in the last year of the study, students were possibly exposed to

© Springer Science+Business Media Singapore and Higher Education Press 2017
J. Zhu, *Understanding Chinese Engineering Doctoral Students in U.S. Institutions*,
East-West Crosscurrents in Higher Education, DOI 10.1007/978-981-10-1136-8_4

project-based learning, teamwork, and other types of educational opportunities (e.g., facing opposing views), which facilitated these students' intellectual development. In the same study, the researchers also found that the first-year design course did demonstrate some positive impact in the beginning and helped with the freshmen's intellectual development. However, it seemed that the possible reason these changes were not maintained could have been because the environment did not support that over the extended period.

Most engineering educators would agree to the concept of moving engineering students to a higher level of intellectual development in which said students would use "the best available evidence to reach conditional acceptance of hypotheses and models and remaining constantly open to reconsideration of conclusions if new evidence is forthcoming" (Felder and Brent 2004, p. 10). This position reflects an acceptance of typical contextual thinking, i.e., Position 5 or higher in Perry's model, and there have been studies on graduate students' intellectual development. Few of these studies, however, are specific to engineering students. It remained unclear whether or not the intellectual development of engineering students will demonstrate the characteristics of the trend that was observed among liberal arts students (Kitchener and King 1981).

4.2 Application of Epistemological Developmental Theories in Graduate Students

For graduate students, Baxter Magolda (1987) did find that doctoral students seemed to exhibit a higher level of thinking than did undergraduate students and master's degree students. In a study she conducted among 39 students (12 freshmen, 10 seniors, 10 master's degree students, and 7 doctoral candidates) from a department of education, she rated the interviews using a score from 1 to 5 indicating Perry's Positions 1–5. For the freshmen, seniors, first-year master's, second-year master's, and doctoral candidates involved in the study, the mean ratings (with standard deviation in parentheses) were 2.61 (0.57), 3.31 (0.58), 3.82 (0.70), 4.13 (0.59), and 4.30 (0.67), respectively. However, she only used Perry's Positions 1–5 and did not explore the students in Perry's Positions 6–9.

It is promising to note that some efforts have been taken to facilitate the development along the lower positions of Perry's scale. But Perry's complete theory has not been used. There have been some other studies that employed Reflective Judgment Interviews (King and Kitchener 1994) to test graduate students' reflective judgment development (King et al. 1990; Kitchener and King 1981). In these instances, it was found that advanced doctoral students demonstrate reflective thinking in King and Kitchener's Reflective Judgment Model. Most of these samples were liberal arts students, however, and the samples used were not engineering student specific. One may speculate that the graduate students would develop further in this transition to become a more relativistic knower, similar to what was indicated in a study on liberal arts students (Kitchener and King 1981); however, no substantial evidence currently exists to support this claim among engineering students.

Nonetheless, in King et al. (1990), they found that graduate students in social science disciplines scored significantly higher on the RJI than did their counterparts in math sciences, suggesting the differences were caused by disciplinary differences. Wankat and Oreovicz (1993) suggested that graduate work in engineering and the physical sciences resembles that of undergraduate work in humanities because they both confront the student with a world of uncertainty and multiplicity. So far, scarce evidence can be found, however, on engineering doctoral students' cognitive development. In sum, more efforts are still needed to understand the cognitive development of engineering graduate students to more accurately understand their particular experiences.

4.3 Application of Epistemological Developmental Theories in Chinese Students

Perry's Theory and other cognitive developmental theories were first developed among Euro-Americans. Several of these theories have been applied to other ethnic groups or countries, such as African American college students (King and Taylor 1989) and college students in Central American, South American (Samson 1999), and Germany (Kitchener and Wood 1987). These studies have found a similar trend among senior students, showing a higher level of cognitive thinking than among the freshmen.

In the past decade, several researchers applied Perry's theory to college students in China (Zhang 1995, 1999, 2000, 2002; Zhang and Hood 1998; Zhang and Watkins 2001). Zhang and her colleagues performed a series of five consecutive studies on the cognitive development of US and Chinese college students over the past decade (see a review of these studies in Zhang 2004). Zhang and her colleagues developed Zhang's Cognitive Development Inventory (ZCDI) based on Perry's theory. It is one of the few survey instruments based on Perry's model.

The latest version (4th version) of the ZCDI was provided to me by Dr. Lifang Zhang through personal correspondence. It has five subscales assessing three of the four positions in Perry's model as mapped in Culver and Hackos (1982). These three positions are *Dualism*, *Relativism*, and *Commitment within Relativism*. *Dualism* and *Relativism* were examined in two content areas, education and interpersonal relationships. *Commitment*, named Life Responsibility in ZCDI, reflects the last position in Perry's model, *Commitment within Relativism*. ZCDI has 75 short statements in all. For each statement, participants provide a response on a 7-point Likert scale indicating the degree to which they agree with the statement.

Zhang (2004) reported internal consistency reliability alpha coefficients ranging from 0.57 to 0.74 for the Chinese sample. The validity of the measure was indicated by the subscale's correlation coefficients fitting their predicted direction. For example, Interpersonal Relationship/Dualism (i.e., Dualism in an Interpersonal Relationship area) is supposed to be inversely correlated with Interpersonal Relationship/Relativism. The reported correlation was $r = -0.32$, $p < 0.01$ for the Chinese sample. Education/Relativism was shown to be positively related with Life Responsibility/Commitment.

They conducted several studies among Chinese college students (193 college students from Nanjing, 1998, and 464 college students from Shanghai, 2004). Based on these studies, they observed a general trend that was the opposite of Perry's developmental trend. They speculated that this reversed developmental trend could be due to the limited opportunities to make choices for Chinese students in terms of curriculum, majors, residential choices, and so on (2004).

Zhang's cognitive development inventory can potentially serve as a tool for applying Perry's theory and understanding the epistemological development within the Chinese population. As related to my research here, I adopted several subscales of the ZCDI, specifically, Education/Dualism, Education/Relativism, and Commitment/ Life Responsibility. In the rest of this dissertation, when referring to the subscales of *Dualism*, *Relativism*, and *Commitment* of ZCDI, I am referring to the subscales of Education/Dualism, Education/Relativism, and Commitment/Life Responsibility, respectively, unless otherwise specified.

It should be pointed out that ZCDI only measures three out of the four stages in Perry's theory, i.e., *Dualism*, *Relativism*, and *Commitment* without *Multiplicity*. Zhang suggested that this difficulty may be due to the fact that students who hold a relativistic view often agree with the statements written for a multiplistic view, too (1995). Moore also observed difficulty in differentiating between these two in their development of a survey (Moore 1989). A careful examination of the positions within *Multiplicity* (P3 and P4) and *Relativism* (P5 and P6) shows some potential reasons for this difficulty.

First of all, Position 4 has two different views within it, specifically, *Multiplicity Correlate* and *Relativism Subordinate*. These two different views are alternatives, i.e., students can potentially take either of them theoretically to get to the next stage, *Relativism*.

Second, it is very important to know that *Multiplicity Correlate* is similar to Position 3 *Multiplicity Subordinate*. This means that a person who holds a view in *Multiplicity Correlate* can exhibit similar thoughts or agree to the statements in *Multiplicity Subordinate*, or vice versa. Meanwhile, for a person with a view in *Relativism Subordinate*, he/she can probably agree with statements written in for someone with a relativistic view, and vice versa.

Therefore, developing an instrument with statements written in Position 4 will be very difficult to differentiate from Position 3 and Position 5. With this in mind, before the actual adoption of ZCDI in the process of quantitative data collection, some modifications were made to the ZCDI subscales with the hope of differentiating *Multiplicity* and *Relativism*. These subscales and the details of modification in this research will be discussed in Chap. 5.

In summary, Zhang and her colleagues developed a pilot survey based on Perry's theory and validated it among both US and Chinese college student populations for the first time. ZCDI can potentially serve as a tool for applying Perry's theory and understanding the epistemological development within the Chinese population. Although their observation about the reversed trend of Chinese college students' developmental trends may or may not hold true for the population in this work—i.e., elite Chinese doctoral students from Chinese universities—their observations do cause one to speculate about the possibilities of making choices and their relation to epistemological development. Given the increased opportunities to make choices in US institutions (curricula, residential choices, etc.), this would be of particular interest in understanding their epistemological development during their doctoral studies in US institutions. The following chapters describe the main research efforts and detailed results. Chapter 5 talks about the modification made to the ZCDI before its actual usage in this research. Chapter 6 delineates the overall research design. Chapters 7 and 8 focus on the overall profiles of students' epistemological development obtained from quantitative data. Chapters 9 and 10 provide details of the stories of the Chinese engineering doctoral students' epistemological thinking styles based on qualitative data.

References

Baxter Magolda, M. B. (1987). Comparing open-ended interviews and standardized measures of intellectual development. *Journal of College Student Personnel, 28*, 443–448.

Culver, R. S., & Hackos, J. T. (1982). Perry's model of intellectual development. *Engineering Education, 72*, 221–226.

Culver, R. S., Woods, D., & Fitch, P. (1990). Gaining professional expertise through design activities. *Engineering Education, 80*, 533–536.

Felder, R. M., & Brent, R. (2004). The intellectual development of science and engineering students. 1. Models and challenges. *Journal of Engineering Education, 93*(4), 269–277.

King, P. M., & Kitchener, K. S. (1994). *Developing reflective judgment: Understanding and promoting intellectual growth and critical thinking in adolescents and adults.* San Francisco: Jossey-Bass.

King, P. M., & Taylor, J. A. (1989). *Intellectual development of Black college students on a predominantly White campus.* Paper presented at the Annual Meeting of the Association for the Study of Higher Education, Atlanta, Georgia.

King, P. M., Wood, P. K., & Mines, R. A. (1990). Critical thinking among college and graduate students. *Review of Higher Education, 13*(2), 167–186.

Kitchener, K. S., & King, P. M. (1981). Reflective judgment: Concepts of justification and their relationship to age and education. *Journal of Applied Developmental Psychology, 2*, 89–116.

Kitchener, K. S., & Wood, P. K. (1987). Development of concepts of justification in German university students. *International Journal of Behavioral Development, 10*, 171–185.

Moore, W. S. (1989). The learning environment preferences: Exploring the construct validity of an objective measure of the Perry scheme of intellectual development. *Journal of College Student Development, 30*, 504–514.

Pavelich, M. J., & Moore, W. S. (1993). Measuring maturation rates of engineering students using the Perry Model. In *Proceedings of the Frontiers in Education Conference,* (pp. 451–455). Washington, DC: American Society for Engineering Education.

Pavelich, M. J., & Moore, W. S. (1996). Measuring the effect of experiential education using the Perry model. *Journal of Engineering Education, 85*(4), 287–292.

Samson, A. W. (1999). *Latino college students and reflective judgment* (Unpublished doctoral dissertation). University of Denver, Colorado.

Wankat, P., & Oreovicz, F. S. (1993). Models of cognitive development: Piaget and Perry. In *Teaching Engineering*. New York: McGraw-Hill. Retrieved from https://engineering.purdue.edu/ChE/AboutUs/Publications/TeachingEng/chapter14.pdf

Wise, J., Lee, S. H., Litzinger, T. A., Marra, R. M., & Palmer, B. (2004). A report on a four-year longitudinal study of intellectual development of engineering undergraduates. *Journal of Adult Development, 11*, 103–110.

Zhang, L. F. (1995). *The construction of a Chinese language cognitive development inventory and its use in a cross-cultural study of the Perry Scheme*. Retrieved from ProQuest: The University of Iowa.

Zhang, L. F. (1999). A comparison of U.S. and Chinese university students' cognitive development: The cross-cultural applicability of Perry's theory. *The Journal of Psychology, 133*(4), 425–439.

Zhang, L. F. (2000). Are thinking styles and personality types related? *Educational Psychology, 20*(3), 271–283.

Zhang, L. F. (2002). Thinking styles and cognitive development. *The Journal of Genetic Psychology, 163*(2), 179–195.

Zhang, L. F. (2004). The Perry scheme: Across cultures, across approaches to the study of human psychology. *Journal of Adult Development, 11*(2), 123–138.

Zhang, L. F., & Hood, A. B. (1998). Cognitive development of students in China and USA: Opposite directions? *Psychological Reports, 82*, 1251–1263.

Zhang, L. F., & Watkins, D. (2001). Cognitive development and student approaches to learning: An investigation of Perry's theory with Chinese and U.S. university students. *Higher Education, 41*, 239–261.

Part II
A Research Design for Measuring Epistemological Development

Chapter 5
Developing a Valid Survey for Personal Epistemology

5.1 Introduction

As discussed in Chap. 2, William Perry's theory can be summarized as four main developmental stages for adult learners, that is, *Dualism*, *Multiplicity*, *Relativism*, and *Commitment within Relativism* (Culver and Hackos 1982), which depict the development of thinking from a dualistic to a relativistic manner. For large sample studies, Zhang and her colleagues developed and tested Zhang's Cognitive Development Inventory (ZCDI) among Chinese college students (Zhang 1995, 1999, 2000, 2002; Zhang and Hood 1998; Zhang and Watkins 2001). Because ZCDI has been validated among Chinese students and was designed to reflect Perry's theory, it was deemed most relevant to this study (Zhu and Cox 2012). However, challenges still exist for measuring the epistemological development in the context of Perry's theory, especially when trying to differentiate the stages of *Multiplicity* and *Relativism*, since no instrument has distinguished these two stages, including ZCDI (Zhang 1995). This challenge is of particular interest because the transition from *Multiplicity* to *Relativism* for a student signifies the most dramatic change in the development of their thinking according to Perry's theory.

This chapter serves as a pre-study for the quantitative data collection and analysis. It focuses on modifying ZCDI by expanding the original subscale of *Relativism* into the subscales of both *Multiplicity* and *Relativism* (Zhu et al. 2015). A survey that includes both the subscales of *Multiplicity* and *Relativism* was constructed from an extensive literature search of current available instruments in the context of Perry's theory. Eight external content experts were invited to participate in a content validity test process. A 22-item survey was finalized via the survey compilation, construction, and testing process. With these processes, the subscales of *Multiplicity* and *Relativism* of this survey expanded the original *Relativism* subscale of ZCDI. The modified ZCDI, which reflects all four stages of Perry's theory, was then used to map Chinese students' epistemological development profile in a larger scale (Chaps. 7 and 8).

© Springer Science+Business Media Singapore and Higher Education Press 2017

J. Zhu, *Understanding Chinese Engineering Doctoral Students in U.S. Institutions*,
East-West Crosscurrents in Higher Education, DOI 10.1007/978-981-10-1136-8_5

5.2 Background

As reviewed in a previous section (Sect. 3.2 Quantitative Measurement of Epistemological Development), quite a few different pen-and-paper measurement methods have been proposed since the first issue of Perry's intellectual and ethical model. These quantitative measures include: Measure of Epistemological Reflection (MER) (Baxter Magolda 1985, 1987, 1992), Learning Environment Preferences (Moore 1989), Reasoning about Current Issues Test (RCI) (King and Kitchener 1994), Survey of Epistemological Beliefs (EBS) (Schommer 1990), and Zhang's Cognitive Development Inventory (ZCDI) (1995, 1997). Based on the comparison of these different quantitative measures that are currently available in the literature (see Sect. 3.2), ZCDI was found to be most relevant to the purpose of applying Perry's theory among the Chinese student population (Zhu and Cox 2012).

Zhang's Cognitive Development Inventory has been piloted and validated among Chinese college students (1995). As is related to my research here, three subscales of the ZCDI, specifically, Education/Dualism, Education/Relativism, and Commitment/Life Responsibility, were adopted in this study. The modifications to ZCDI are based upon these three subscales, which reflect three of the four stages of Perry's theory, i.e., *Dualism*, *Relativism*, and *Commitment*.

Zhang's Cognitive Development Inventory effectively allows researchers to understand the Chinese students' epistemological development profile, although it only reflects three out of the four stages. It has not been found to differentiate *Multiplicity* and *Relativism* because of the difficulties in item construction and the possible overlap between these two stages of thinking (Zhang 1995). A similar difficulty in differentiating *Multiplicity* and *Relativism* was also observed in Fago's study (1995). Fago constructed a 45-item survey from Perry's original description and applied the survey among 751 college students. Through a factor analysis of the survey results, he observed only three factors, which again reflect only three of the four stages in Perry's model—*Dualism, Relativism,* and *Commitment*.

These are several reasons for the difficulty in these studies in differentiating between *Multiplicity* and *Relativism*. In Perry's original discussions (Perry 1970), Position 3 (*Multiplicity Subordinate*) denotes an incomplete state of *Multiplicity*. There are two alternative pathways within Position 4, i.e., Position 4a-*Multiplicity Correlate* and Position 4b-*Relativism Subordinate*, each representing an alternative pathway to true relativistic thinking. Position 4a-*Multiplicity Correlate* is cognitively closer to Position 3-*Multiplicity Subordinate* in that people in these two positions exhibit different levels of multiplistic thinking. It is therefore reasonable to group them in the measurement for the stage of *Multiplicity*. However, a person in Position 4b-*Relativism Subordinate* may show some similar thinking or behavioral patterns to a person in Position 5-*Relativism*. In Position 5 (*Relativism*), people would analyze, weigh, and evaluate different factors, causes, and arguments. In Position 4b (*Relativism Subordinate*), a person may try to mimic a way of reasoning similar to that of Position 5 because this is the way of thinking professors or other authoritative figures want him/her to have or "the way They want you to think"

(Perry 1970, p. 112). Therefore, grouping Position 3 and Positions 4a and 4b together under *Multiplicity* can potentially complicate the measurement.

The author speculates that because of the similarity between Position 3-*Multiplicity Subordinate* and Position 4a-*Multiplicity Correlate*, survey items that reflect these two positions would probably be hard to differentiate from each other and therefore should be grouped together. Meanwhile, because of the similarities between Position 4b-*Relativism Subordinate* and Position 5-*Relativism*, the items will tend to overlap each other.

With the arguments listed above, within this chapter, statements were designed or compiled into a survey to specifically reflect Positions 3, 4, and 5. The statements for the two alternative views within Position 4, i.e., *Multiplicity Correlate* and *Relativism Subordinate,* were also separated. With these efforts, the goal is to be able to separate the views of *Multiplicity* and *Relativism*. To accomplish this goal, a detailed content validity check was conducted to examine the extent to which every statement within the survey reflects the positions within *Multiplicity* and *Relativism*, i.e., Position 3-*Multiplicity Subordinate*, Position 4a-*Multiplicity Correlate*, Position 4b-*Relativism Subordinate*, and Position 5-*Relativism*. The results from the content validity check were used to separate the subscales of *Multiplicity* and *Relativism*. Also, structural validation was accomplished through an exploratory factor analysis using responses from 621 Chinese engineering students to examine the structure of the resultant survey.

5.3   Survey Construction

In order to differentiate Perry's stages of *Multiplicity* and *Relativism* via a pen-and-paper measurement tool, an extensive literature search was conducted concerning the available pen-and-paper measurement tools for mapping adult's epistemological development within the context of Perry's theory. Built upon the literature search, especially on current available instruments, a 24-item survey was constructed to reflect the four specific positions within the Perry's theory, i.e., (1) Position 3-*Multiplicity Subordinate*, (2) Position 4a-*Multiplicity Correlate*, (3) Position 4b-*Relativism Subordinate*, and (4) Position 5-*Relativism*, which are the positions within the stages of *Multiplicity* and *Relativism*.

Due to the subtle differences across the stages and the positions within these stages, it is challenging to design items that can be categorized clearly into each stage. Therefore, the focus here is to compile, revise, or construct items that can clearly reflect the different positions within the two stages across a large sea of current literature. Specifically, items were compiled from several instruments (Fago 1995; Moore 1989; Zhang 1997) that have items that are relevant to the views of *Multiplicity* and *Relativism*. Also some new items were constructed based on the description of the thinking at different positions according to Perry's narration (Perry 1970).

Items were beta-tested among six Chinese graduate students from different majors in a large Midwestern university to make sure the meaning of each item is clear to the Chinese students. Editorial changes were made. Based upon an item's original design along the Perry's scale and an understanding of the current literature in Perry's theory, each item was grouped into the positions of Perry's theory. The list of survey items, along with their original literature source and their Perry positions, is shown in Table 5.1. The ones that have a citation of Perry (1970) are items composed according to Perry's original descriptions within specific positions.

Table 5.1 Survey items compiled and constructed from an extensive literature review (Zhu et al. 2015)

Perry positions[a]	Items	Source
P3/P4-a	Where authorities differ, a student's opinion should be graded only on how well it is expressed	Perry (1970)[b]
P3/P4-a	I think it's unfair that instructors give a better grade to one student's answer than the other's answer, even when there is no clear answer for the problem	Fago (1995)
P3/P4-a	It doesn't seem fair when grades aren't proportional to work efforts. Many times a person can receive a better grade on a paper that he hasn't worked hard on than a paper that he really worked on	Fago (1995)
P4-a	If you really get deeper into the material in a particular field, what you find is that nobody understands it	Fago (1995)
P4-a	In areas where there is no one correct viewpoint but multiple viewpoints, I often feel unclear as to the standards that instructors use to evaluate students' viewpoints	Perry (1970)
P4-a	When experts in a particular field disagree with one another, no one really knows the answer	Perry (1970)
P4-a	In issues where experts have no consensus, it can be confusing, because there is no way to prove whether one viewpoint is more reasonable than the other	Perry (1970)
P4-a	In an academic debate, there are merits in both sides' views; therefore, the criteria to decide which side wins are not clear	Perry (1970)
P4-a	I am certain of one thing—even if there is an absolute truth, we will never know about it, and, therefore, there is no correct answer to most questions	Zhang (1997)
P4-a	Since people's views are influenced by their educational backgrounds, almost any view could be as valid as any other	Zhang (1997)
P4-b	It seems to me that some instructors try to get you to look at something in a complex way by weighing multiple factors at once	Fago (1995)
P4-b	I try to think in an independent manner, because thoughts that appear to be independent get good grades	Perry (1970)
P4-b/P5	When writing an open-ended essay, I take a stance after thinking about other possible viewpoints	Zhang (1997)

(continued)

Table 5.1   (continued)

Perry positions[a]	Items	Source
P4-b/P5	Taking a stand during an academic debate requires reasoning and taking risks	Zhang (1997)
P5	I find I can detach myself emotionally from problems and look at their various sides in order to formulate a judgment	Fago (1995)
P5	I enjoy classes with a seminar format where students can exchange their ideas so I can critique my own perspectives on the subject matter	Moore (1989)
P5	I prefer to be in a class that is loosely structured where the students take most of the responsibility for the structure of the class	Moore (1989)
P5	I enjoy a class where opportunities are provided for me to pull together connections among various subject areas and then construct an adequate argument	Moore (1989)
P5	I would like opportunities to think on my own and make connections between the issues discussed in class and in other areas I'm studying	Moore (1989)
P5	When I solve a problem, I often think through several alternatives to find the best solution	Perry (1970)
P5	It is very hard for me to accept a teacher's view on controversial issues when he/she does not provide enough evidence to support his/her view	Perry (1970)
P5	It is not difficult for me to give up ideas and opinions I hold if I find that my classmates' ideas sound more reasonable	Zhang (1997)
P5	Integrating ideas is my favorite part of writing a paper	Zhang (1997)
P5	I enjoy working with complex ideas in which experts have no consensus	Zhang (1997)

Note
[a]Perry positions: P3-*Multiplicity Subordinate*; P4a-*Multiplicity Correlate*; P4b-*Relativism Subordinate*; P5-*Relativism*
[b]This is an item developed based on the Checklist of Educational Values, an instrument that Perry (1970) used in his original sampling procedure

5.4   Content Validation

The goal is to differentiate the stages of thinking between *Multiplicity* and *Relativism*. Because differences across the stages and the positions within are delicate and subtle, it is challenging to design items that are specific to the two stages. Content validity check was a procedure commonly used to examine the degree to which an instrument reflects the guiding theoretical framework in its design of survey items (Carmines and Zeller 1991). Here, by inviting experts who are familiar with this framework, one can potentially differentiate these two stages, i.e., *Multiplicity* and *Relativism*.

In addition, the subtle differences across the positions within the two stages could potentially complicate the process of differentiating the two stages in some other statistical analysis, such as factor analysis. Therefore, it is more reasonable to first perform the content validity check at the individual position level, i.e., Position 3-*Multiplicity Subordinate*; Position 4a-*Multiplicity Correlate*; Position 4b-*Relativism Subordinate*; Position 5-*Relativism*, in order to first parse the subtle differences across the positions and to obtain meaningful data.

5.4.1 Data Collection and Analysis

To check the content validity of the survey, eight experts were invited to rate the extent to which each survey item reflected the following positions of Perry's theory: Position 3-*Multiplicity Subordinate*; Position 4a-*Multiplicity Correlate*; Position 4b-*Relativism Subordinate*; Position 5-*Relativism*. These external experts represent a variety of perspectives and experiences with Perry's theory. According to their self-reports, they have either engaged in research efforts related to Perry's theory, taught some classes related to Perry's theory, or attended training sessions (classes, workshops, etc.) related to Perry's theory. One of the experts self-indicated that he had led numerous workshops on the Perry scheme and had been active in research on the scheme for many years.

Table 5.2 presents a sample item of the Expert Rating Form as presented to content experts. All experts rated each item along Positions 3 to 5 using a Likert scale from one to four (1, not at all; 2, only a little; 3, somewhat, 4, a great deal). An open-ended question was also added to collect any additional comments to the survey items. An information page was provided to the experts with brief definitions of each position within Perry's theory. The information page and the Expert Rating Form can be found in Appendix A. After the experts rated all of the items, the ratings across content experts for each position of each item were compiled. If the sum of external experts rating for a certain position of an item is higher than 24,

Table 5.2 A sample item on the expert rating form

Directions: Please rate the degree to which the statement reflects the view of Perry positions (P) 3, 4a, 4b, and 5 using the following scale: 1: not at all, 2: only a little, 3: somewhat, 4: a great deal

Survey item	P3-Multiplicity subordinate	P4a-Multiplicity correlate	P4b-Relativism subordinate	P5-*Relativism*
	1 2 3 4	1 2 3 4	1 2 3 4	1 2 3 4
E.g., Everyone has a right to his/her own opinion. Anything goes	☐ ☒ ☐ ☐	☐ ☐ ☒ ☐	☒ ☐ ☐ ☐	☒ ☐ ☐ ☐

i.e., the average score among all eight experts is equal to or higher than 3 (somewhat), then the item will be rated as in the corresponding position. In order to identify students' epistemological developmental stage, it is preferable that each item only reflect one particular position or possibly its neighboring positions. Therefore, the items that reflect two unrelated positions will be deemed as not useful. For example, items that had high scores on both *Multiplicity Subordinate* and *Relativism* are deemed not fitting.

5.4.2 Results

Through the process of the content validity check, each survey item was determined in terms of the degree to which it reflected the four positions within Perry's stages of *Multiplicity* and *Relativism*, i.e., Position 3-*Multiplicity Subordinate*; Position 4a-*Multiplicity Correlate*; Position 4b-*Relativism Subordinate*; Position 5-*Relativism*. The sum of the ratings for each item in each position from all of the eight experts was compiled. The results for all the items in each of the four positions are shown in Table 5.3.

Based upon the content validation results, 20 out of the 24 survey items were identified as representing at least one of the four positions of Perry's theory (see the shaded numbers in Table 5.3). Among the 20 items, the classifications of Perry's positions for 18 items matched the originally designated classifications according to literature review and the author's understanding. This overall result demonstrates a high level of design validity of the survey items to accomplish their original purposes of reflecting a certain position of Perry's theory.

By a detailed examination of the individual item, it appears that that the items reflecting the Position 3-*Multiplicity Subordinate* and/or the Position 4a-*Multiplicity Correlate*, i.e., items 1 to 10, are very close in meaning and therefore have quite a lot of overlap. For instance, items 2 and 7 reflect both Position 3-*Multiplicity Subordinate* and Position 4a-*Multiplicity Correlate*. The ratings of some other items also suggest this similarity, for example, items 1 and 3 both have quite high rating in both Position 3 and Position 4a (higher than 20). Items within Position 3-*Multiplicity Subordinate* and Position 4a-*Multiplicity Correlate* mostly reflect the students' awareness of the diversity of views and their dilemma in deciding the criteria against which these views can be evaluated when "no correct answer is available." It is reasonable to conclude that the classifications of these items from external ratings suggested the similarity of Position 3-*Multiplicity Subordinate* and Position 4a-*Multiplicity Correlate* in their reflections of the stage of *Multiplicity*.

Built upon these findings, the items that reflect either P3 or P4a or both were grouped as items of *Multiplicity* in the final classifications. It should be noted that this classification validates the conventional way of classifying *Multiplicity* in the literature (Culver and Hackos 1982), that is, Position 3 *Multiplicity Subordinate* and Position 4a-*Multiplicity Correlate* were both in the stage of *Multiplicity*. Although, in the conventional way of classifying *Multiplicity*, Position 4b-*Relativism*

Table 5.3 Grand total of rating for each survey item across experts (Zhu et al. 2015)

Survey items		Original Perry positions	Perry positions (experts' rating)				Final classification
			P3-Multiplicity subordinate	P4a-Multiplicity correlate	P4b-Relativism subordinate	P5-Relativism	
1	Where authorities differ, a student's opinion should be graded only on how well it is expressed	P3/P4-a	31	21	17	12	*Multiplicity*
2	I think it's unfair that instructors give a better grade to one student's answer than the other's answer, even when there is no clear answer for the problem	P3/P4-a	29	25	15	9	*Multiplicity*
3	It doesn't seem fair when grades aren't proportional to work efforts. Many times a person can receive a better grade on a paper that he hasn't worked hard on than a paper that he really worked on	P3/P4-a	27	22	17	10	*Multiplicity*
4	If you really get deeper into the material in a particular field, what you find is that nobody understands it	P4-a	11	25	18	15	*Multiplicity*
5	In areas where there is no one correct viewpoint but multiple viewpoints, I often feel unclear as to the standards that instructors use to evaluate students' viewpoints	P4-a	31	19	14	10	*Multiplicity*
6	When experts in a particular field disagree with one another, no one really knows the answer	P4-a	22	23	14	11	*Multiplicity*

(continued)

Table 5.3 (continued)

Survey items	Original Perry positions	Perry positions (experts' rating)				Final classification
		P3-*Multiplicity subordinate*	P4a-*Multiplicity correlate*	P4b-*Relativism subordinate*	P5-*Relativism*	
7	In issues where experts have no consensus, it can be confusing, because there is no way to prove whether one viewpoint is more reasonable than the other	P4-a	28	24	17	9	*Multiplicity*
8	In an academic debate, there are merits in both sides' views, therefore, the criteria to decide which side wins are not clear	P4-a	22	25	21	14	*Multiplicity*
9	I am certain of one thing—even if there is an absolute truth, we will never know about it and, therefore, there is no correct answer to most questions	P4-a	13	21	14	24	Not Included
10	Since people's views are influenced by their educational backgrounds, almost any view could be as valid as any other	P4-a	12	27	19	14	*Multiplicity*
11	It seems to me that some instructors try to get you to look at something in a complex way by weighing multiple factors at once	P4-b	7	13	22	22	*Relativism*
12	I try to think in an independent manner, because thoughts that appear to be independent get good grades	P4-b	18	17	21	17	Not Included

(continued)

Table 5.3 (continued)

Survey items		Original Perry positions	Perry positions (experts' rating)					Final classification
			P3-*Multiplicity subordinate*	P4a-*Multiplicity correlate*	P4b-*Relativism subordinate*	P5-*Relativism*		
13	When writing an open-ended essay, I take a stance after thinking about other possible viewpoints	P4-b/P5	8	11	20	31		*Relativism*
14	Taking a stand during an academic debate requires reasoning and taking risks	P4-b/P5	11	14	22	28		*Relativism*
15	I find I can detach myself emotionally from problems and look at their various sides in order to formulate a judgment	P5	9	11	18	28		*Relativism*
16	I enjoy classes with a seminar format where students can exchange their ideas so I can critique my own perspectives on the subject matter	P5	8	12	18	32		*Relativism*
17	I prefer to be in a class that is loosely structured where the students take most of the responsibility for the structure of the class	P5	8	17	17	25		*Relativism*
18	I enjoy a class where opportunities are provided for me to pull together connections among various subject areas and then construct an adequate argument	P5	8	9	17	31		*Relativism*

(continued)

Table 5.3 (continued)

Survey items		Original Perry positions	Perry positions (experts' rating)				Final classification
			P3-*Multiplicity subordinate*	P4a-*Multiplicity correlate*	P4b-*Relativism subordinate*	P5-*Relativism*	
19	I would like opportunities to think on my own and make connections between the issues discussed in class and in other areas I'm studying	P5	8	12	15	31	*Relativism*
20	When I solve a problem, I often think through several alternatives to find the best solution	P5	12	14	20	31	*Relativism*
21	It is very hard for me to accept a teacher's view on controversial issues when he/she does not provide enough evidence to support his/her view	P5	12	14	20	31	*Relativism*
22	It is not difficult for me to give up ideas and opinions I hold if I find that my classmates' ideas sound more reasonable	P5	10	13	22	22	*Relativism*
23	Integrating ideas is my favorite part of writing a paper	P5	8	13	20	29	*Relativism*
24	I enjoy working with complex ideas in which experts have no consensus	P5	11	16	16	26	*Relativism*

Subordinate was also included in the *Multiplicity*. This difference will be explained as follows.

In contrast to the conventional way of classifying Position 4b-*Relativism Subordinate* in the *Multiplicity* in current literature (Culver and Hackos 1982), the ratings from the external experts seemed to suggest a similarity between Position 4b-*Relativism Subordinate* and Position 5-*Relativism* in their conceptual representation. Items that were designed for Position 4b-*Relativism Subordinate* and Position 5- *Relativism* were rated to be similar in their representations of Perry positions. Item 11, 13, 14, 20, 21, 22, and 23 all have quite high ratings in both Position 4b and Position 5 (equal to or higher than 20). This observation raised a concern about the conventional way of putting Position 4, including both 4a and 4b, into the stage of *Multiplicity*. Conceptually, the essence of Position 4b *Relativism Subordinate* seems to be closer to P5-Relativism than P3 *Multiplicity Subordinate* and Position 4a-*Multiplicity*. In this sense, classifying both Positions 4a and 4b within *Multiplicity* can complicate the differentiation between *Multiplicity* and *Relativism*, which could potentially cause a difficulty in the measurement of these ways of thinking.

Several items (i.e., items 6, 11, and 12) did not satisfy the criteria of obtaining a total score higher than 24. Explanations about these items are mentioned below.

For item 6, "When experts in a particular field disagree with one another, no one really knows the answer," the ratings for Position 3-*Multiplicity Subordinate* and the Position 4a-*Multiplicity Correlate* were 22 and 23, respectively, which strongly indicates that the item represents the stage of *Multiplicity*. Therefore, it was still *included* in the final survey.

For item 11, "It seems to me that some instructors try to get you to look at something in a complex way by weighing multiple factors at once," the ratings for Position 4b-*Relativism Subordinate* and Position 5-*Relativism* were both 22, which indicates that the item represents the stage of *Relativism*. Therefore, the item was also *included* in the final version of survey.

For item 12, "I try to think in an independent manner, because thoughts that appear to be independent get good grades," the ratings for Position 3, Position-4a, Position-4b, and Position-5 are 18, 17, 21, and 17 respectively. Because of the low scores across each position for this item, it did not reflect any position in a consistent manner. Therefore, it was *excluded* from the final version of the survey.

Finally, for item 9, "I am certain of one thing—even if there is an absolute truth, we will never know about it and, therefore, there is no correct answer to most questions," it also has a sum larger than 24 in Position 5-*Relativism*. However, it was originally classified as Position 4a-*Multiplicity Correlate*. With the rating of 21 in Position 4a-*Multiplicity Correlate* and 24 in Position 5-*Relativism* from experts' rating, this item is excluded from the final version of the survey for the purpose of differentiating *Multiplicity* and *Relativism*.

To summarize, a 22-item survey was constructed and validated via the survey construction based on an extensive literature review and external content experts' ratings (see the Final Classification column in Table 5.3). Using this survey, along with the final classification in *Multiplicity* or *Relativism*, one can potentially differentiate students' thinking along these two stages. By combining this survey with current available surveys, here in this study, with *Dualism/Education* and *Commitment/Life Responsibility* subscales from the ZCDI (Zhang 1997), one can possibly conduct a large-scale study for the understanding of the epistemological development of Chinese students in the context of Perry's theory.

5.5   Structural Validation

We adopted the survey that was obtained through the content validity check in this section. An exploratory factor analysis was conducted to explore the structure of this current survey. Exploratory factor analysis is an important tool to examine structural validity in the early stages of an instrument development. Its goal is to "identify the underlying dimensions of a domain of functioning, as assessed by a particular measuring instrument" (Floyd and Widaman 1995). Here, I further explored the potential construct(s) within the survey and compare the constructs with results from experts' ratings.

5.5.1   Data Collection and Analysis

Six hundred twenty-one responses were collected from engineering undergraduate and graduate students who majored in 17 engineering disciplines such as Mechanical Engineering, Electrical Engineering, and Materials Science and Engineering in a leading Chinese university. These survey responses were then used for exploratory factor analysis.

Principal component analysis (PCA) was used to extract factors. Overlap can exist among these factors; therefore, an oblique rotation (Promax) was employed to explain them. Specifically, the survey obtained from the content validity check was administered after a few minor wording changes. A five-point Likert scale was used (strongly disagree, disagree, not sure, agree, strongly agree) to record students' responses to these short statements.

5.5.2   Results

Considering the research goal of differentiating *Multiplicity* and *Relativism*, a two-factor model was identified as the most suitable model with the Kaiser-Meyer-Olkin (KMO) value (0.786) indicating acceptable sampling adequacy (Kaiser 1974). Factor loadings of all the items in this model are listed in Table 5.4. Factor loadings that were larger than 0.35 were considered as loaded to a certain factor (Zhang 1995).

Table 5.4 Item mapping for a two-factor model (Zhu et al. 2015)

Item	Factor loading		Original Perry position	Experts' rating
	1	2		
Factor one				
1. In an academic debate, there are merits in both sides' views; therefore, the criteria to decide which side wins are not clear	0.133	**0.584**	P4-a	M
2. I think it's unfair that instructors give a better grade to one student's answer than the other's answer, even when there is no clear answer for the problem	0.067	**0.565**	P3/P4-a	M
3. In issues where experts have no consensus, it can be confusing, because there is no way to prove whether one viewpoint is more reasonable than the other	0.043	**0.547**	P4-a	M
4. If you really get deeper into the material in a particular field, what you find is that nobody understands it	0.058	**0.532**	P4-a	M
5. In areas where there is no one correct viewpoint but multiple viewpoints, I often feel unclear as to the standards that instructors use to evaluate students' viewpoints	0.082	**0.459**	P4-a	M
6. It doesn't seem fair when grades aren't proportional to work efforts. Many times a person can receive a better grade on a paper that he hasn't worked hard on than a paper that he really worked on	0.125	**0.427**	P3/P4-a	M
7. When experts in a particular field disagree with one another, no one really knows the answer	−0.318	**0.423**	P4-a	M
8. Where authorities differ, a student's opinion should be graded only on how well it is expressed	0.152	**0.355**	P3/P4-a	M
Factor Two				
1. I find it helpful to be in a class where opportunities are provided for me to pull together connections among various subject areas and then construct an adequate argument	**0.661**	−0.013	P5	R
2. I would like opportunities to think on my own and make connections between the issues discussed in class and in other areas I'm studying	**0.644**	0.004	P5	R
3. I prefer to be in a class where the instructor engages students to contribute to the structure of the class	**0.586**	−0.050	P5	R

(continued)

Table 5.4 (continued)

Item	Factor loading		Original Perry position	Experts' rating
	1	2		
4. When I solve a problem, I often think through several alternatives to find the best solution	**0.584**	−0.006	P5	R
5. I prefer classes with a seminar format where students can exchange their ideas so I can critique my own perspectives on the subject matter	**0.576**	0.024	P5	R
6. Integrating ideas is my favorite part of/on writing a paper	**0.521**	0.078	P5	R
7. It seems to me that some instructors try to get you to look at something in a complex way by weighing multiple factors at once	**0.478**	0.206	P4-b	R
8. When writing an open-ended essay, I take a stance after thinking about other possible viewpoints	**0.435**	-0.051	P4-b/P5	R
9. Taking a stand during an academic debate requires reasoning and taking risks	**0.429**	0.161	P4-b/P5	R
10. I find I can detach myself emotionally from problems and look at their various sides in order to formulate a judgment	**0.402**	0.109	P5	R
11. It is very hard for me to accept a teacher's view on controversial issues when he/she does not provide enough evidence to support his/her view	**0.380**	0.208	P5	R
12. Since people's views are influenced by their educational backgrounds, almost any view could be as valid as any other	**0.350**	0.154	P4-a	M
Other				
1. I enjoy working with complex ideas in which experts have no consensus	0.334	0.165	P5	R
2. It is not difficult for me to give up ideas and opinions I hold if I find that my classmates' ideas sound more reasonable	0.272	0.097	P5	R

Abbreviations: M, *Multiplicity*; R, *Relativism*
Factor loadings > 0.35 appear in bold

In all, eight items were loaded to factor one. This factor seems to reflect mostly the dimension of *Multiplicity*. Meanwhile, 12 items were loaded to factor two, which seems to represent the dimension of *Relativism* except for one item. Based upon the loading, it is reasonable to conclude that the structure of the survey has already reflected the two dimensions of Perry's theory. Also, two items were not loaded to any factors.

As shown in Table 5.4, items 2, 6, and 8 in factor one, with the original Perry positions as P3/P4a, were found to load onto factor one, which seems to representing *Multiplicity* in which most items are of the original Perry positions P4a. This finding suggests the similarity between P3-*Multiplicity Subordinate* and Position 4a-*Multiplicity Correlate*. Item 7 in factor two, with original Perry position as P4b, loaded onto the same factor representing *Relativism* as items whose original Perry positions were P4b/P5 or P5. This finding suggests the similarity between Position 4b-*Relativism Subordinate* and P5-*Relativism*.

In summary, through a content validity check and exploratory factor analysis, we confirmed that this survey can reflect the two dimensions of *Multiplicity* and *Relativism* in Perry's theory. Using this survey, one can differentiate students' thinking along these two stages. Combined with current available measure, this survey can help map students' epistemological development stages in the context of Perry's theory.

5.6   Discussion

The results from the external content experts' rating suggest that the items that reflect Position 3- *Multiplicity Subordinate* and Position 4a-*Multiplicity Correlate* tend to overlap with each other. This is probably due the similarities in the behavioral patterns of individuals who are in these two positions. Meanwhile, because of the similarities of the behavioral patterns of individuals who are in Position 4b- *Relativism Subordinate* and Position 5-*Relativism*, the items under these two positions tend to overlap with each other. Based upon the content validity test results, a 22-item survey was finalized with 9 items reflecting *Multiplicity* and 13 items reflecting *Relativism*.

Based upon the results from the content validity study, it is suggested that instead of using the conventional way of grouping Position 3-*Multiplicity Subordinate*, Position 4a-*Multiplicity Correlate,* and Position 4b-*Relativism Subordinate* together as *Multiplicity*, it is more reasonable to separate Position 4a and Position 4b because of the similarities between Position 4b *Relativism Subordinate* and Position 5-*Relativism*. By so doing, it is then more likely to separate *Multiplicity* and *Relativism* conceptually and therefore produce a valid instrument that can differentiate the thinking between *Multiplicity* and *Relativism*.

By applying this survey, one can separate individuals with the view of *Multiplicity* (including *Multiplicity Subordinate* and *Correlate*) from individuals with the view of *Relativism* (including *Relativism Subordinate* and *Relativism*). In the literature, several currently available surveys have proven to successfully measure the other two stages of Perry's theory, i.e., *Dualism* and *Commitment within Relativism* (Zhang 1995; Fago 1995), but not the stages of *Multiplicity* and *Relativism*. Therefore, the differentiation between *Multiplicity* and *Relativism* fills a current gap of literature in measuring Perry's developmental stages using quantitative measures.

Moreover, this survey tool was compiled and constructed for specific applications among the Chinese population. In the compilation and construction processes, the applicability among Chinese students has been taken into consideration. Also, the survey was beta-tested among the Chinese population for clarity and understanding. Therefore, the survey is readily applicable for measuring the epistemological development of Chinese students. Nonetheless, the content of the survey were designed based on current literature and author's understanding of Perry's theory. The overall content should also be applicable across different contexts. With some additional testing and changes, it is expected that this survey can be applied among other ethnic groups.

One limitation for this instrument is that the survey cannot help differentiate the views of *Relativism Subordinate* and *Relativism*. It should be noted that only a few items in the final survey reflect Position 4b-*Relativism Subordinate*. This is partially due to the difficulty to design items in Position 4b *Relativism Subordinate*. The limitation can potentially be addressed by in-depth follow-up interviews to understand the motivations of the behavioral patterns that are mimicking relativistic thinking.

5.7 Conclusion

This chapter reports the construction, content, and structural validation of a survey modified based on Zhang's Cognitive Development Inventory and other currently available instruments in the context of Perry's theory. It provides a way to potentially separate students who are in the stages of *Multiplicity* and *Relativism*. Incorporating this instrument and the Dualism/Education and Commitment/Life Responsibility subscales of the ZCDI, the modified ZCDI provides a way to measure and understand individuals' thinking along the four stages of Perry's scale (*Dualism, Multiplicity, Relativism,* and *Commitment*).

This section serves as a pre-study for the larger study of mapping the cognitive development profile of Chinese doctoral students in engineering using both quantitative and qualitative methods. The modified ZCDI was applied to a large sample of Chinese engineering doctoral students across five Midwestern universities to map students' epistemological development (Chaps. 7 and 8), and follow-up interviews were conducted to provide an in-depth picture of students' epistemological development (Chaps. 9 and 10).

References

Baxter Magolda, M. B. (1985). A new approach to assessing intellectual development on the Perry scheme. *Journal of College Student Personnel, 26,* 343–351.

Baxter Magolda, M. B. (1987). Comparing open-ended interviews and standardized measures of intellectual development. *Journal of College Student Personnel, 28,* 443–448.

Baxter Magolda, M. B. (1992). *Knowing and reasoning in college*. San Francisco: Jossey-Bass.

Carmines, E. G., & Zeller, R. A. (1991). *Reliability and validity assessment*. Newbury Park: Sage Publications.

Culver, R. S., & Hackos, J. T. (1982). Perry's model of intellectual development. *Engineering Education, 72*, 221–226.

Fago, G. C. (1995). A scale of cognitive development: Validating Perry's scheme.

Floyd, F. J., & Widaman, K. F. (1995). Factor analysis in the development and refinement of clinical assessment instruments. *Psychological Assessment, 7*(3), 286–299.

Kaiser, H. F. (1974). An index of factorial simplicity. *Psychometrika, 39*(1), 31–36.

King, P. M., & Kitchener, K. S. (1994). *Developing reflective judgment: Understanding and promoting intellectual growth and critical thinking in adolescents and adults*. San Francisco: Jossey-Bass.

Moore, W. S. (1989). The learning environment preferences: Exploring the construct validity of an objective measure of the Perry scheme of intellectual development. *Journal of College Student Development, 30*, 504–514.

Perry, W. G. (1970). *Forms of intellectual and ethical development in the college years: A scheme*. New York: Holt, Rinehart and Winston.

Schommer, M. (1990). Effects of beliefs about the nature of knowledge on comprehension. *Journal of Educational Psychology, 82*(3), 498–504.

Zhang, L. F. (1995). *The construction of a Chinese language cognitive development inventory and its use in a cross-cultural study of the Perry scheme*. Retrieved from ProQuest: The University of Iowa.

Zhang, L. F. (1997). *The Zhang cognitive development inventory* (Unpublished text). The University of Hong Kong, Hong Kong.

Zhang, L. F. (1999). A comparison of U.S. and Chinese university students' cognitive development: The cross-cultural applicability of Perry's theory. *The Journal of Psychology, 133*(4), 425–439.

Zhang, L. F. (2000). Are thinking styles and personality types related? *Educational Psychology, 20*(3), 271–283.

Zhang, L. F. (2002). Thinking styles and cognitive development. *The Journal of Genetic Psychology, 163*(2), 179–195.

Zhang, L. F., & Hood, A. B. (1998). Cognitive development of students in China and USA: Opposite directions? *Psychological Reports, 82*, 1251–1263.

Zhang, L. F., & Watkins, D. (2001). Cognitive development and student approaches to learning: An investigation of Perry's theory with Chinese and U.S. university students. *Higher Education, 41*, 239–261.

Zhu, J., & Cox, M. F. (2012). Epistemological development of Chinese engineering doctoral students in the U.S. institutions: A comparison of multiple measurement methods. In *2012 Proceedings of the American Society for Engineering Education*.

Zhu, J., Hu, Y., Liu, Q., & Cox, M. F. (2015). Validation of an instrument for Chinese engineering students' epistemological development. *International Journal of Chinese Education, 4*, 135–161.

Chapter 6
A Mixed-Method Research Design

Using the survey developed in Chap. 5, this chapter illustrates the motivation and the process of using an explanatory mixed-method research design for understanding students' epistemological development. In this work, Perry's theory was applied for studying the epistemological development of Chinese engineering doctoral students who are pursuing doctorate degrees in US institutions. Through the use of an explanatory mixed-methods based research design (Creswell 2008), both quantitative and qualitative data were collected among Chinese engineering doctoral students from five Big Ten universities; the purposes of this research are:

1. *To identify the epistemological development profiles of Chinese engineering doctoral students and*
2. *To understand the factors related to the epistemological development profiles among Chinese engineering doctoral students.*

Overarching research questions to be considered include the following:

1. *What are the epistemological development profiles of Chinese engineering doctoral students framed within the context of the modified Perry theory on epistemological development?*
2. *Based on the research results of research question 1, what are the possible factors that are related to these profiles?*

To attain the research goals and objectives, an explanatory mixed methods strategy was adopted in the research design (Creswell 2008). In this research design, quantitative and qualitative data were collected sequentially. As Creswell stated, this methodological design captures *"the best of both quantitative and qualitative data—to obtain quantitative results from a population in the first place"* and *"refine or elaborate these findings through an in-depth qualitative exploration in the second phase"* (Creswell 2008, p. 560).

© Springer Science+Business Media Singapore and Higher Education Press 2017
J. Zhu, *Understanding Chinese Engineering Doctoral Students in U.S. Institutions*,
East-West Crosscurrents in Higher Education, DOI 10.1007/978-981-10-1136-8_6

Built upon the pre-study of survey validation in Chap. 5, this research was composed of the following two major studies:

Study 1 Quantitative Profiles of Chinese Engineering Doctoral Students' Epistemological Development

After the modifications to the ZCDI, quantitative data were collected using the revised survey carried out among 147 Chinese engineering doctoral students from five Big Ten universities. A subscale from another widely adopted survey, the Epistemological Belief Survey (Wood and Kardash 2002), was also used for the purpose of data triangulation. A demographic survey was administered among the participants to collect information about different variables. Quantitative data collection and results provided an overview of the epistemological development profiles of Chinese engineering doctoral students. Results from quantitative data collection provided a base and guidance for the procedures of the qualitative data collection and analysis.

Study 2 Stories of Chinese Engineering Doctoral Students' Epistemological Thinking Styles

Informed from the results of the second study, 19 students from the survey respondents were engaged in one-on-one interviews to understand the participants' lived experiences. The one-on-one interviews were guided by the phenomenological

Fig. 6.1 Research design

methodological framework, which is the main methodology recommended for understanding personal epistemology, and were adopted in nearly all of the major studies in young adults' epistemological development (Perry 1970; Belenky et al. 1986; Baxter Magolda 1992). Through the collection and analysis of qualitative data, the epistemological thinking styles of Chinese engineering doctoral students were operationalized in the context of Perry's theory. Factors that are associated with said students' epistemological development were also explored. The findings from qualitative data collection and analysis were used to confirm and refine the results of the second study.

The overall flow of the research plans and steps is shown in Fig. 6.1. Combined results from both the quantitative and qualitative data provided a better understanding of the epistemological development profiles of Chinese doctoral students and the necessary details as to the potential factors that are related with the epistemological development of Chinese engineering doctoral students.

References

Baxter Magolda, M. B. (1992). *Knowing and reasoning in college*. San Francisco: Jossey-Bass.
Belenky, M. F., Clinchy, B. M., Goldberger, N. R., & Tarule, J. M. (1986). *Women's ways of knowing: The development of self, voice and mind*. New York: Basic Books.
Creswell, J. W. (2008). *Research design*. London: Sage Publications.
Perry, W. G. (1970). *Forms of intellectual and ethical development in the college years: A scheme*. New York: Holt, Rinehart and Winston.
Wood, P., & Kardash, C. (2002). Critical elements in the design and analysis of studies of epistemology. In B. K. Hofer & P. R. Pintrich (Eds.), *Personal epistemology: The psychology of beliefs about knowledge and knowing* (pp. 231–260). Mahwah, NJ: Erlbaum.

Part III
A Quantitative Exploration of Chinese Engineering Doctoral Students' Epistemological Development

Chapter 7
Overall Profiles of Epistemological Development

The purpose of this study is to understand the overall profiles of the Chinese engineering doctoral students' epistemological developmental status through use of Perry's scale, i.e., the measurement of *Dualism, Multiplicity, Relativism*, and *Commitment*. In this chapter, I shall describe the details of the quantitative data collection, data analysis, and interpretation. The results gathered from quantitative data collection and analysis provide an overall picture of the Chinese engineering doctoral students' epistemological development as viewed through the lens of Perry's theory. It will also serve as the basis for qualitative data collection and analysis.

7.1 Background

Although limited past research has been performed on the epistemological development of *Chinese engineering doctoral* students, current research does provide some useful information on the epistemological development of the *engineering* students, *doctoral* students, and *Chinese* students, respectively. Understanding these student groups' epistemological development can render insight into the investigation of the target population.

With regards to engineering students, Pavelich and Moore (1996) have found that only one fourth of the graduating engineering seniors had developed up to Position 5 (Relativism) with an overall average rating for graduating seniors at 4.28 ± 0.70 (Position 4 is the Multiplicity Correlate or Relativism Subordinate). This finding was later confirmed by Wise et al. (2004) with a similar range of 4.2 ± 0.50 for seniors. Although their studies did not specifically include graduate level engineering students, their findings suggest that high level thinking (i.e., equal to or higher than Position 5 Relativism) exists among the graduate engineering students within the US graduate program.

© Springer Science+Business Media Singapore and Higher Education Press 2017
J. Zhu, *Understanding Chinese Engineering Doctoral Students in U.S. Institutions*,
East-West Crosscurrents in Higher Education, DOI 10.1007/978-981-10-1136-8_7

Moreover, for graduate students, Baxter Magolda (1987) found that doctoral students seemed to show a higher level of thinking than did undergraduate students and master's degree students in a study among 39 students in the School of Education. It is a pity that she only used Perry's Position 1–5 for rating the interviews and did not explore the students' representation in Perry's Positions 6–9. I suspect that some of the doctoral students studied could have already exhibited Perry's higher positions (i.e., equal to or higher than Position 5).

For the Chinese students, only those at the college level were explored in terms of their epistemological development. In Zhang and colleagues' studies among Chinese college students' epistemological development (Zhang 1995, 1999, 2000, 2002; Zhang and Hood 1998; Zhang and Watkins 2001), they found a reversed developmental trend to the one described in Perry's theory. She noted that the students in upper classes (senior and junior) seemed to score higher on the dualistic thinking subscale and lower on the relativistic thinking subscale and the *Commitment* subscale. She suspected that this trend could result from the lack of choices available for majors, curricula, dormitory options, and so on, in Chinese colleges. Zhang's series of studies has raised an interesting concern about the relationship that exists between available choices and the students' epistemological development.

Informed from current literature concerning the target population, the *Chinese engineering doctoral* students who are studying in *US graduate programs*, several aspects require further attention to better understand said students' epistemological development.

Although no direct research has been conducted to study Chinese engineering doctoral students quantitatively in terms of their epistemological development, the current literature still provides some useful background information for this study. The purpose in this particular study is to explore the epistemological developmental profiles of Chinese engineering doctoral students and to explore factors that are associated with the profiles in a quantitative manner. The research questions of this dissertation are:

1. *What are the epistemological development profiles of Chinese engineering doctoral students framed within the context of the modified Perry theory on epistemological development?*
2. *Based on the research results of research question 1, what are the possible factors that are related to these profiles?*

The goal of this study is to explore the overall profiles of the Chinese engineering doctoral students' epistemological development. It should be noted that the term "profile" refers to a snapshot of developmental stages at a given point for a certain student population. Therefore, it provides a cross-section of the student population at the specific time of the experiment instead of providing longitudinal information about these students. Later, the qualitative data and the analysis may also provide some deeper insight into the terms of the students' retrospective understanding of their own experiences. The quantitative study does not provide this longitudinal information.

To accomplish the goal of the quantitative study, a survey was employed. This survey includes three major parts. The first part is the modified ZCDI from Chap. 5. This is the main method by which to map the epistemological development profiles of Chinese engineering doctoral students in context of Perry's theory. The second part is the subscale of Knowledge Construction and Modification from the Epistemological Beliefs Survey (EBS) (Wood and Kardash 2002). It is used together with the revised ZCDI to serve the purpose of triangulation for data collection. The third part is a demographic survey.

The survey was distributed to approximately 1,000 Chinese engineering doctoral students in the fall of 2012 across five Midwestern universities. Overall, 147 complete responses were collected. The following sections describes the process of data collection and defines the major findings. The results based on the quantitative data provide an overview of the epistemological development profiles of Chinese engineering doctoral students.

7.2 Method

7.2.1 Data Collection

Survey

This survey includes three major parts: (1) the Modified Zhang's Cognitive Developmental Inventory (48 items); (2) the subscale of Knowledge Construction and Modification from EBS (11 items); (3) demographic questions (14 questions). The whole set of surveys were created using the Qualtrics survey software via the Purdue University website. The compiled survey with the first and second part can be found in Appendix B. Demographic questions can be found in Appendix C. Each portion of the survey is described in further detail as follows.

Modified Zhang's Cognitive Developmental Inventory (ZCDI)

The original ZCDI was developed based on Perry's framework. It measures *three* of the four positions that were traditionally acknowledged as the main stages in his framework (Culver and Hackos 1982). These four positions are *Dualism, Multiplicity, Relativism,* and *Commitment (within Relativism)*. The three positions measured by the ZCDI are *Dualism, Relativism,* and *Commitment (within Relativism)*. In particular, the items in the subscale of *Relativism* in the ZCDI actually reflect both *Multiplicity* and *Relativism* and were not able to differentiate between these two via measurement. Based on the work described in Chap. 5, the two subscales were separated via the content experts' ratings to represent a more holistic picture of Perry's theory. By revising the subscale of *Relativism* into two subscales, *Multiplicity* and *Relativism*, the modified ZCDI used in this research can be seen to represent all four of Perry's stages. The items within these four subscales add up to 48 items. Out of the 48 items, 20 were from the Education/Dualism subscale and 7 from the Commitment subscale of the ZCDI. Another 21 items were from the survey obtained via Chap. 5. One item in the 22-item survey obtained from Chap. 5 was not included because it is similar to an item in the Knowledge Construction and Modification subscale from the EBS.

Subscale of Epistemological Belief Survey (EBS)

In the EBS, the Knowledge Construction and Modification factor relates closely to the participants' epistemic development from a dualistic view to a more constructivist view. As described by Wood and Kardash (2002), "High scores on this factor (i.e. the Knowledge Construction and Modification factor) reflect the ideas that knowledge is ... personally constructed. By contrast, low scores on this factor reflect a view that knowledge is certain, passively received, and accepted at face value" (p. 250). Therefore, this factor appears to be a fairly good indication of the students' epistemological development in terms of their abilities in knowledge construction and modification. This subscale was used for this study as an additional data collection method for the purpose of triangulation. Higher scores in the scale reflect that a student's idea is leaning toward the thinking that knowledge is personally constructed. In contrast, lower scores in the scale indicate that a student's perception about knowledge is leaning towards an understanding that knowledge is certain and passively received.

Demographic Questions

Demographic questions were compiled from multiple sources to collect different variables. Several items were adapted from the background section of the Survey of Earned Doctorates (SED 2012), which is designed to gather information from doctorate graduates. A question about geographic origins in China was modified from a survey from a Chinese data source (MYCOS 2010 report, p. 60). Another question about the parents' educational backgrounds in China was modified from a demographic survey by Zhang (1995, p. 289). Other elements were added to reflect key elements of graduate studies, such as one's academic progress in graduate school, undergraduate institutions, whether one holds a master's degree or not, engineering disciplines, prior work experiences, and so on.

Sampling

Chinese doctoral students majoring in engineering disciplines in the Midwest US were recruited for this study. Appropriate Institutional Review Board procedures were followed before any quantitative survey was sent out to the various students' email lists.

Considering the quantitative nature involved in this part of study, a large sample pool was required for accurate statistical computations. Nationwide, out of the top 25 institutions hosting international students in 2009/10, 9 of these were the Big Ten universities (Table 7.1). The top five Big Ten universities in the list are: University of Illinois-Urbana-Champaign, Purdue University-Main Campus, University of Michigan-Ann Arbor, Michigan State University, and Ohio State University. The number of Chinese graduate students for each university is shown in Table 7.2.

The five Big Ten universities are all public universities with both a large enrollment and very active research activities. They are located primarily within the Midwestern area. The universities' enrollment of total international students and Chinese graduate students is more or less comparable with regards to their actual numbers. In the presentation of data in the later section, the names of the universities will be replaced by pseudonyms. Pseudonyms, University A, B, C, D, and E, were randomly assigned to these universities.

The survey described above was distributed to Chinese engineering doctoral students across these five universities. Specifically, only those students who fully

Table 7.1 Top 15 institutions hosting international students 2009/10 (IIE report 2010, p. 12)

Rank	Institution	City	State	Total int'l students	Total enrollment
1	University of Southern California	Los Angeles	CA	7987	34,824
2	University of Illinois-Urbana-Champaign	Champaign	IL	7287	43,723
3	New York University	New York	NY	7276	43,208
4	Purdue University-Main Campus	West Lafayette	IN	6903	41,051
5	Columbia University	New York	NY	6833	24,188
6	University of Michigan-Ann Arbor	Ann Arbor	MI	6095	41,674
7	University of California-Los Angeles	Los Angeles	CA	5685	39,750
8	Michigan State University	East Lansing	MI	5358	47,278
9	University of Texas-Austin	Austin	TX	5265	50,995
10	Boston University	Boston	MA	5172	31,499
11	University of Florida	Gainesville	FL	4920	50,691
12	SUNY University at Buffalo	Buffalo	NY	4911	28,881
13	Harvard University	Cambridge	MA	4867	26,500
14	Indiana University-Bloomington[a]	Bloomington	IN	4819	42,347
15	Ohio State University-Main Campus	Columbus	OH	4796	55,014

[a]Indiana University-Bloomington did not have an engineering school

Table 7.2 Top five universities with most international students among the Big Ten universities

Institution	Total int'l students	Chinese graduate students	Year
University of Illinois-Urbana-Champaign	8648	1694	2012[a]
Purdue University-Main Campus	8562	1179	2012[b]
University of Michigan-Ann Arbor	8491	N/A	2012[c]
Michigan State University	6599	788	2012[d]
Ohio State University	4796	N/A	2010 (IIE report 2010)

[a]http://www.dmi.illinois.edu/stuenr/index.asp#foreign
[b]https://www.iss.purdue.edu/ISSOffice/Reports.cfm, The number 1179 also include some professional students in addition to graduate students
[c]www.internationalcenter.umich.edu/Annual_Report.pdf
[d]http://oiss.isp.msu.edu/documents/statsreport/12pdfs/CountryLevel.pdf

satisfied these four criteria received the recruitment email with a link to the survey. These four criteria are: (1) study level: graduate students; (2) nationality: international students from PR China (including students from Hong Kong and Macao, but not from Taiwan); (3) college: from the college of engineering; (4) degree objective: doctorate degree (PhD). Emails were sent via the help of the specific departments that are responsible for these types of requests from these universities (e.g., the registrar's office or department of information management).

After the recruitment, 147 responses were received. A breakdown of recruitment emails sent to the different universities and responses collected from these universities is shown in Table 7.3.

The overall response rate is 12 %. Considering the length of the survey (59 items plus 14 demographic questions), a response rate lower than 20 % is common according to the current literature (Marcus et al. 2007).

A more detailed distribution of the sample population, sorted by age, gender, marital status, number of children, place of origin, area of living, religion, parent's level of education, the type of undergraduate institution attended ("211" vs.

Table 7.3 Distributions of responses collected among different universities

Institution	Number of recruitment emails sent	Number of responses
University A	291	40
University B	256	34
University C	261	24
University E	260	24
University D	153	7
Unknown	0	18
Total	1,221	147

"non-211"), prior master's degree (yes/no), academic progress, and work experience, is shown in Figs. 7.1, 7.2, 7.3, 7.4, 7.5, 7.6, 7.7, 7.8, 7.9, 7.10, 7.11 and 7.12.

As shown in Fig. 7.1, approximately half of the students are not older than 25 years of age. For the rest of the students, most are not over 30 years old. Only 2 % are over 30 years of age. These percentages suggest few of the study subjects had had extensive work experience before they joined their doctoral programs.

As shown in Fig. 7.2, 74 % of the respondents are male, and 26 % are female. With regard to marital status, over half of the respondents are single (Fig. 7.3). Only 29 % are married, and 7 % are living in a committed relationship. Corresponding to this trend, fewer than 30 % of the respondents have children who are under 18 years of age within their households (Fig. 7.4). The rest of these respondents do not have children.

One of the demographic questions surveyed the province from which the participants originated. According to the students' responses, it seems that their places of origin are quite diverse. Using a classification modified from a Chinese government report (Liu 2005), there are students from each of the first eight regions (see Fig. 7.5). As for the specific areas from which the students originated, nearly

Fig. 7.1 The distribution of survey respondents by age

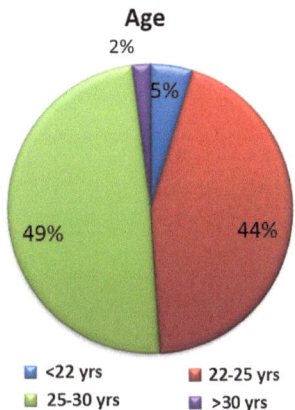

Fig. 7.2 The distribution of survey respondents by gender

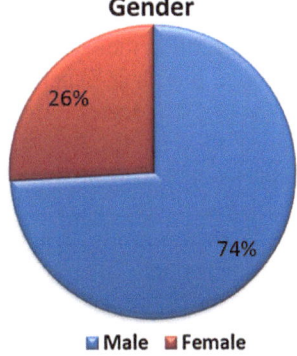

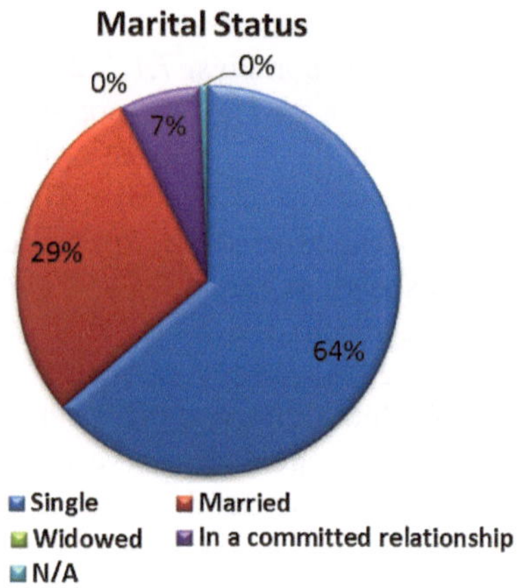

Fig. 7.3 The distribution of survey respondents by marital status

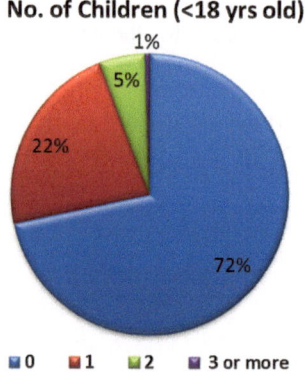

Marital Status

- Single
- Married
- Widowed
- In a committed relationship
- N/A

Fig. 7.4 The distribution of survey respondents by the number of children (<18 years old)

No. of Children (<18 yrs old)

- 0
- 1
- 2
- 3 or more

70 % are from urban areas, 21 % are from suburban areas, and 10 % are from rural areas (Fig. 7.6).

One of the demographic questions asked was regarding the student's religion. According to the students' responses (Fig. 7.7), 38 % did not provide this information, 27 % of the students indicated that they are atheists, 17 % said they had no religion, 10 % are Christians, and the rest indicated other religions.

Concerning their parents' level of education (Fig. 7.8), over half of the students reported that their fathers had some college experience or held a bachelor's, master's, or PhD degree. A slightly reduced percentage of collegiate education was found to exist for the students' mothers. Still, more than half of the students reported that their mothers had had some college experience or held a bachelor's, master's, or PhD degree.

As shown in Fig. 7.9, the majority of the students' reported the undergraduate institutions are "211" type (see archives from the Chinese Ministry of Education

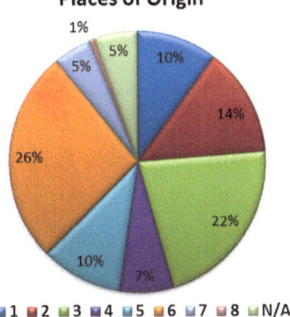

Fig. 7.5 The distribution of survey respondents by their places of origin. *Note 1* Northeast district (Liaoning, Jilin, and Heilongjiang). *2* North coastal district (Beijing, Tianjing, Hebei, and Shandong). *3* East coastal district (Shanghai, Jiangsu, and Zhejiang). *4* Southeast coastal district (Guangdong, Fujian, and Hainan). *5* Upper and middle reaches of the Yellow River (Shanxi, Gansu, Ningxia, Shaanxi, and Henan). *6* Upper and middle reaches of the Yangtze River (Sichuan, Chongqing, Hubei, Hunan, Anhui, and Jiangxi). *7* Upper and middle reaches of the Pearl River (Yunnan, Guizhou, and Guangxi). *8* Inner Mongolia district. *9* Xinjiang district. *10* Qingzang Altiplano district (Qinghai and Xizang). (No participants are from the last two of the ten areas)

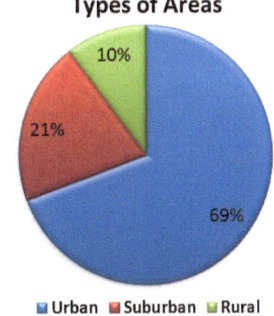

Fig. 7.6 Distributions of survey respondents by area where they live

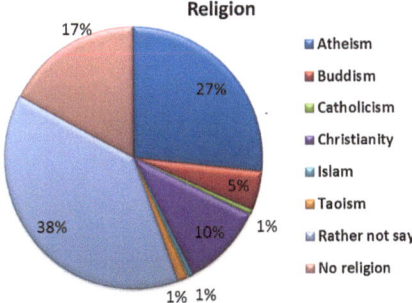

Fig. 7.7 The distribution of survey respondents by religion

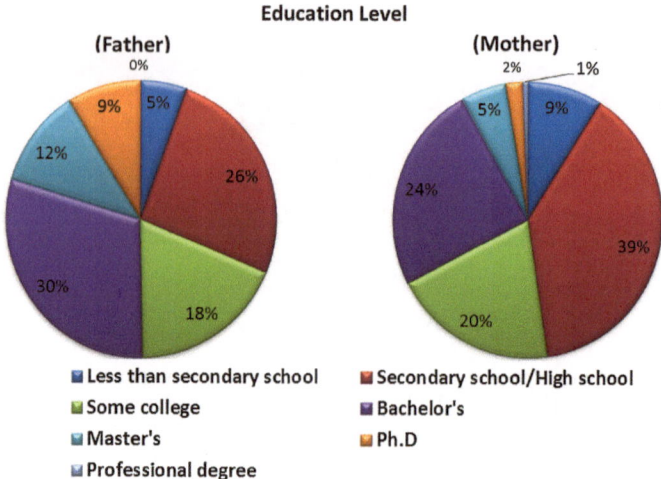

Fig. 7.8 Distributions of survey respondents by parents' education level

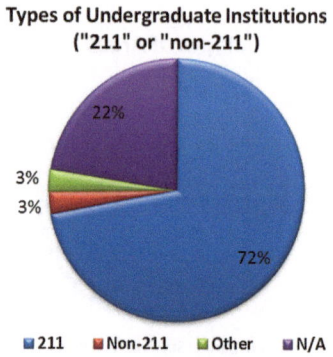

Fig. 7.9 Distributions of survey respondents by the type of their undergraduate institutions. *Note* "Other" refers to institutions outside of the classification system by the Ministry of Education in China

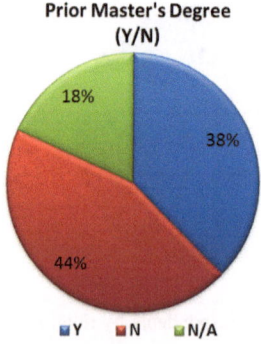

Fig. 7.10 Distributions of survey respondents by whether the respondent has a master's degree or not

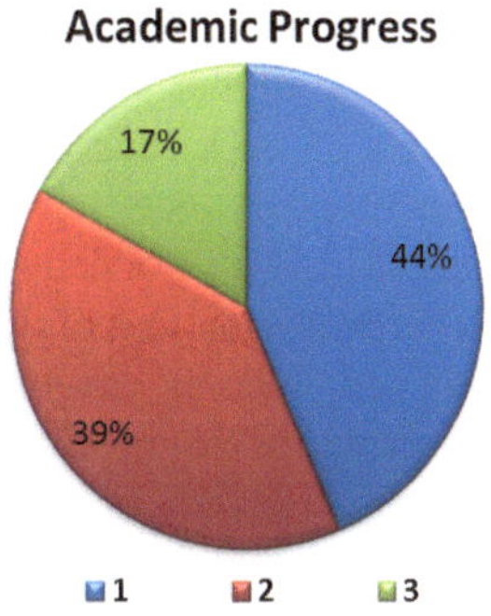

Fig. 7.11 Distributions of
survey respondents by
academic progress. *Note 1*
course work stage; *2* passed
qualifying and/or preliminary
examinations or other similar
milestone examinations; *3*
dissertation stage

Academic Progress

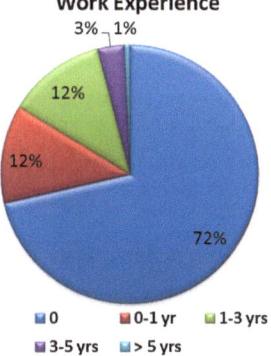

Fig. 7.12 Distributions of
the sample population by
prior work experience

Work Experience

website, 2009). About 22 % of the students did not provide information about their
undergraduate institutions. Only 3 % are from "non-211" type institutions. The
final 3 % attended other institutions outside of the classification system of the
Ministry of Education in China, such as universities in Hong Kong, or other
countries and regions, such as universities in Singapore, North America, and so on.
As illustrated in Fig. 7.10, 44 % indicated that they have obtained a master's
degree, 38 % did not have a master's degree, and the rest of the studied population
did not provide this information.

As is illustrated in Fig. 7.11, 44 % of the students reported that they were only
working on their course work; 39 % of the students reported that they have passed
their qualifying, preliminary, or other equivalent candidacy examinations. This
study involves students from varied academic programs, which have differing
policies regarding these examinations. Some programs use the preliminary examinations as the candidacy examination, while other programs use a qualifying
examination as the candidacy examination. Therefore, these different examinations
are categorized together into one group. Finally, 17 % of students reported that they
were at the dissertation research stage of their studies.

As is shown in Fig. 7.12, 72 % of the students have no prior work experience, while 12 % of the students have 0–1 years of experience. Another 12 % have 1–3 years of work experience, 3 % have 3–5 years experience, and 1 % have over 5 years of experience.

Out of 147 respondents, 79 did not provide information about their current engineering major. Out of the respondents who provided the information about their current engineering majors, 26 students are from Electrical and Computer Engineering, 8 from Mechanical Engineering, 6 from Industrial Engineering, 6 from Materials Sciences and Engineering, 5 from Biomedical Engineering, 4 from Aeronautics and Astronautics Engineering, 3 from Civil Engineering, 2 from Chemical Engineering, and 8 from other types of engineering majors.

7.2.2   Data Analysis

The survey responses for this study were downloaded through the Qualtrics software of the Purdue University website. The responses were analyzed using statistical procedures described below.

T-Test

T-Test procedures were conducted to evaluate the scores within each subscale for each student. Specifically, the average scores for all of the items within each subscale (i.e., *Dualism*, *Multiplicity*, *Relativism*, and *Commitment*) were calculated for each student. For any student, if the average score for a subscale is higher than 3, i.e., neutral, in a statistically significant way, then the dimension(s) in which the average score is higher than 3 is established to be the student's stage of cognitive development. For example, if a student's average score for *Dualism* is greater than 3 in a statistically significant way, while the average scores for all of the other dimensions are not larger than 3 in a statistically significant way, then the student's determined cognitive developmental stage is *Dualism*. In the case where two or more than two subscales have average scores larger than 3 in a statistically significant way, then the subscale(s) that has an average score larger than 3 with highest statistical significance level is determined as the student's cognitive developmental stage.

In statistical hypothesis testing, to determine whether the average score of a subscale for a student is higher than 3, the hypothesis test can be expressed via this equation,

$$T = \frac{(\bar{X} - 3)}{s/\sqrt{n}}$$

In this equation, T refers to the test statistic for this calculation; $\bar{X}$ represents the average scores of a subscale; s refers to the standard deviation of the scores within that subscale; n refers to the number of items in that subscale.

Therefore, the hypotheses are as follows:

Null hypothesis $H_0 : \bar{X} \leq 3$;

Alternative hypothesis $H_a : \bar{X} > 3$;

The test statistic T follows T-distribution.

The students may have variations in their individual discerning of a high score versus a low score. That is to say, students may perceive each term of rating (e.g., "strongly agree") differently from one another. Therefore, different significance levels were tried for all of the students. When the significance level (α) was set at the level of 0.001, the cognitive stages of 64 students from the studied population of 147 respondents were determined. That is to say, the confidence level is 99.9 %. When the significance level (α) is set at the level of 0.005, the cognitive stages of 28 students out of the remaining population were determined. That is to say, the confidence level is 99.5 %. When the significance level (α) is set at the 0.01 level, the cognitive stages of 15 students out of the remaining population were determined. That is to say, the confidence level is 99 %. When the significance level α is set at the level of 0.025, the cognitive stages of 11 students out of the remaining population were determined. When the significance level is set at the level of 0.05, where the confidence level is lowered to 95 %, the cognitive stages of 11 students out of the remaining population were determined. When the significance level is set at the 0.10 level, where the confidence level is lowered to 95 %, the cognitive stages of another 8 students out of the remaining population were determined. Any cognitive stages that are shown with a confidence level that is lower than 90 % (i.e., significance level α higher than 0.10) are not taken into consideration for the purpose of grouping students into their cognitive stages. In all, 137 students' cognitive developmental stages were identified using this method.

The Knowledge Construction and Modification subscale from the EBS, as an additional subscale, was also used in the survey to provide triangulation in data collection. Therefore, in the overall data analysis process, survey results of this subscale are also provided for complementary information.

ANOVA

One-way analysis of variance (ANOVA) procedures were used to test for significant difference in the four subscales separately (*Dualism, Multiplicity, Relativism, and Commitment*) (dependent variables) with regards to different demographic parameters (independent variables). These independent variables include age, gender, marital status, number of children, place of origin, area of living, religion, parent's educational level, prior master's degree (yes/no), current university, academic progress, work experience, and the students' current universities. For example, one-way ANOVA was used to determine the difference in the *Dualism* subscale with regards to varying years of work experiences. The Scheffé procedure was used for all of the post hoc comparisons.

It should be noted that only the results that indicate significant differences are shown in the results section. No significant differences were observed for these variables: age, gender, marital status, number of children, area of living, religion, parent's educational level, and work experience. Significant differences were observed in some subscales for these variables: academic progress, place of origin, prior master's degree (yes/no), and the students' current universities. Details are discussed in the next chapter.

7.3   An Overall Profile of Epistemological Stages

Through use of the combined survey of the modified ZCDI, the Knowledge Construction and Modification subscale from the EBS, and the demographic survey, the goal of this quantitative survey and the analysis is to provide an overall epistemological developmental profile of the Chinese engineering doctoral students and to explore potential factors that are related with the profile.

Using the above-mentioned T-test procedures, the epistemological stages for all of the participants were determined by identifying the most prominent thinking across the four dimensions in Perry's theory. The epistemological stages of 137 students (out of 147) were identified using the T-test procedures with a significance level alpha of no more than 0.10, i.e., confidence level higher than 90 %. The epistemological stages or the most prominent thinking styles among the 137 students are shown in Fig. 7.13. The distribution of the number of students in each different group is provided in Fig. 7.14.

Several observations can be summarized from the presence of different epistemological stages in this sample and the overall distribution of students across different groups (Figs. 7.13 and 7.14).

First, nearly half of the students fall into the Relativism group. That is to say, these engineering doctoral students' prominent epistemological thinking is relativistic thinking. According to Perry's theory, the development to relativistic thinking is the most significant transition in one's epistemological development. To reflect relativistic thinking signifies that one individual has begun to adopt relativistic reasoning to evaluate different information. This observation shows that relativistic thinking is present among the Chinese engineering doctoral students in the US graduate programs. In addition, the finding suggests that almost half of the students have already developed on to the stage of Relativism in their epistemological thinking.

Second, over 80 % of the students fall into the groups of *Relativism, Relativism-Commitment*, and *Commitment (within Relativism)*. This finding suggests that the prominent thinking of over 80 % of the students has demonstrated some format of relativistic thinking. Moreover, 30 % of the students have already exhibited a commitment to relativistic thinking. To exhibit a commitment to relativistic thinking indicates one individual has begun to take responsibilities in some major areas of his/her life, such as one's career choices (Perry 1970). This finding suggests that most of the students have demonstrated higher levels of thinking in Perry's theory (i.e., relativistic thinking and/or commitment to relativistic thinking).

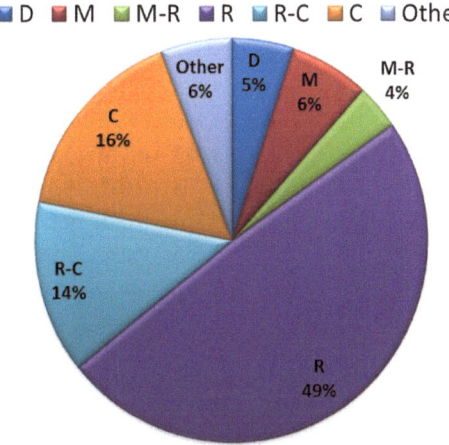

Fig. 7.13 Different groups present in the Chinese engineering doctoral student sample (*n* = 137) using students' prominent epistemological thinking as a grouping method. *Note* "Other" includes D–R, M–R–C, D–C, and D–M, with fewer than three persons in each group. Abbreviations: D, *Dualism*; M, *Multiplicity*; R, *Relativism*; C, *Commitment*

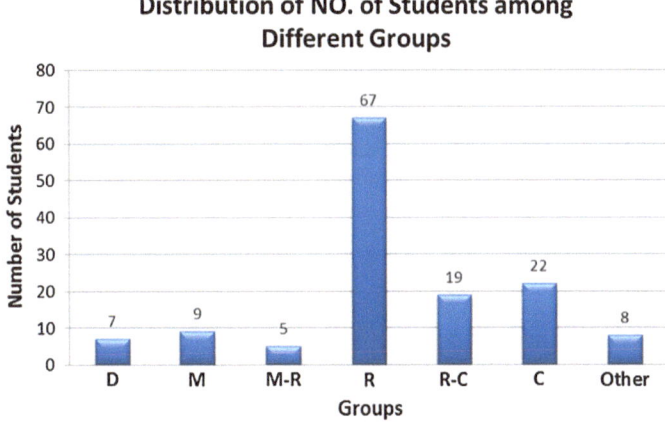

Fig. 7.14 Distribution of the number of students among different groups. Abbreviations: D, *Dualism*; M, *Multiplicity*; R, *Relativism*; C, *Commitment*

Third, around 20 % of students demonstrate two or more of Perry's stages at the same time. This finding suggests that students' epistemological developmental stage can be very complicated with multiple epistemological thinking styles present at the same time. It is possible that some of these students within the sample might

Table 7.4 Summary of the average scores for the students in each group

Groups	n	Subscales									
		Dualism		Multiplicity		Relativism		Commitment		EBS-KCM	
		Ave.	Std. error	Ave.	Std. error	Ave.	Std. error	Ave.	Std. error	Ave.	Std. error
D	7	3.657	0.145	3.492	0.153	3.345	0.100	3.184	0.147	3.649	0.173
M	9	2.694	0.128	3.667	0.135	3.250	0.089	3.143	0.130	3.616	0.153
M–R	5	2.990	0.172	3.667	0.181	3.683	0.119	3.400	0.174	4.073	0.205
R	67	2.701	0.047	3.123	0.050	3.795	0.032	3.358	0.048	4.019	0.056
R–C	19	2.634	0.088	3.006	0.093	3.816	0.061	4.015	0.089	4.139	0.105
C	22	2.655	0.082	3.030	0.086	3.455	0.057	3.987	0.083	4.021	0.098

Note The survey scores are defined as follows: *1* strongly disagree; *2* disagree; *3* neutral; *4* agree; *5* strongly agree

Abbreviations: D, *Dualism*; M, *Multiplicity*; R, *Relativism*; C, *Commitment*

be going through some transitional stage where nearby epistemological developmental stages are present concurrently.

A summary of the average scores for the students in each group is shown in Table 7.4. The average score for the Knowledge Construction and Modification (KCM) subscale from the EBS is also provided for the purpose of data triangulation. Theoretically speaking, higher KCM average scores would indicate stronger ability in constructive or relativistic manners of thought. Numerically, the data suggest that the average scores of the KCM subscale for the groups of *Dualism* and *Multiplicity* are probably lower than those of the groups of *Relativism*, *Relativism-Commitment*, and *Commitment*. Although the significance level for this difference is only about 0.146 and does not guarantee a statistically significant difference, considering the overall trend, the results from the modified ZCDI and the KCM subscale in the EBS do agree with each other in the scoring of these students' epistemic thinking.

7.4 Discussion

To summarize, the overall profile obtained from the quantitative survey results shows that the prominent thinking of a major portion of the Chinese engineering doctoral students is relativistic thinking (over 50 %). Most of the students' prominent thinking styles fell into the higher level of thinking (i.e., *Relativism* or *Commitment*) according to Perry's theory (nearly 80 %), with only 15 % or so of the students with prominent thinking styles that fell into the lower levels of Perry's theory (i.e., *Dualism* or *Multiplicity*). This finding provides a snapshot of the epistemological development profile of Chinese engineering doctoral students in US graduate programs. This finding fills the gap in the current research on the epistemology in the following aspects.

First, Pavelich and Moore (1996) have suggested that only a quarter of the graduating engineering seniors have developed to a position above Position 5, which is Relativism. Also, Baxter Magolda (1987) measured graduate and under-graduate education students' epistemological development in a large Midwestern US university. Her results suggested an average rating of 4.30 ± 0.67 for doctoral candidates, a measurement that indicates that some students showed relativistic thinking skills in their epistemological thinking development. No such research results, however, have been provided directly with regards to engineering doctoral students. For the first time, this study focuses on engineering doctoral students and helps the understanding of the students' epistemological development from an engineering discipline-specific perspective. The method and results of this study among Chinese students can also serve as an example to future similar studies among other ethnic groups to permit a better understanding of the engineering doctoral students' epistemological development.

Second, Zhang and her colleagues found a trend for Chinese college students' epistemological development that was opposite to the trend described in Perry's scale (Zhang 1995, 1999, 2000, 2002; Zhang and Hood 1998; Zhang and Watkins 2001). She suspected that the opposite trend could be associated with the lack of choices available in majors, curricula, dormitory options, and so on, within the Chinese collegiate system. No direct research, however, had been conducted to explore the Chinese students after college. Therefore, this study provides first-hand quantitative evidence for the presence of relativistic thinking among Chinese doctoral students within US engineering programs. These results may provide useful information for further studies on Chinese students in other graduate pro-grams outside of the US systems, e.g., students in Chinese graduate schools or graduate programs in other countries. Research designs including these different Chinese populations can potentially assist in understanding the factors evident in different educational systems that are possibly related with students' epistemolog-ical development.

The overall profile of the Chinese engineering doctoral students' epistemological development indicates that nearly 80 % of the students exhibit relativistic thinking and/or commitment (in relativistic thinking). Several factors are possibly associated with this distribution.

First of all, as stated in Sect. 1.1, only 1–2 % of Chinese college graduates migrated to the US to pursue science and engineering degrees. Most of these students were primarily high achievers from 211 type universities (MYCOS 2010). Therefore, the level of epistemological thinking of these students could potentially be higher than an average score of all students, such as the one reported in Zhang's prior research (Zhang 2004).

Second, in the process of migrating to the US and while studying in the US graduate programs, these students encounter many changes both in the culture and the educational system, in which they face different culture norms and/or academic expectations (Wang 2009). Theoretically, the exposure to a variety of views and opinions could potentially facilitate one's development toward a higher level of thinking regarding measurement according to Perry's theory (i.e., relativistic thinking or commitment to relativistic thinking, Perry 1970). Moreover, the US doctoral engineering education features the development of independent research skills, critical thinking skills, innovative skills, teamwork skills, communication skills, and so on (Golde and Walker 2006; Cox et al. 2011). The development of many of these skills can also facilitate one's epistemological development.

Third, all of the prior studies focused on Chinese college students (Zhang 1995, 1999, 2000, 2002; Zhang and Hood 1998; Zhang and Watkins 2001) were completed over a decade ago. During the last two decades, several national projects have been launched by the Chinese Ministry of Education to enhance both the educational quality and academic and research standards in a number of major universities within Chinese higher education. The two most important projects have been the "211 Project" (1995) and the "985 Project" (1998) (see archives from the Chinese Ministry of Education website 2008). The goals of these projects are to produce high-level talents, solve major scientific and technological problems in economic development, and focus the limited funds on the major universities with solid infrastructures and key disciplines. With the implementation of these educational projects, rich educational resources have been provided to support key universities in China, which has since then enhanced the teaching and research quality of these universities. Many of the target Chinese engineering students are from 211 institutions (MYCOS 2010); therefore, the quality of students, including their epistemological development, could potentially be quite different from the student quality ten years ago.

7.5 Conclusion

To summarize, the overall profile fills a gap in the current understanding of young adults' epistemological development by focusing on Chinese engineering doctoral students. The profile of Chinese engineering doctoral students' epistemological development can also offer practical guidance for the sampling, interview process, and other aspects of the qualitative data collection and analysis.

In addition to the implications to the studies in epistemological development, findings from the current studies can inform and facilitate the qualitative data collection and analysis in the following different ways:

First, the overall profiles of Chinese engineering doctoral students' epistemological development can provide a direction for the qualitative research effort. That is, most of the students' prominent thinking styles fell into the groups of *Relativism* or *Commitment* in Perry's theory (nearly 80 %). This major finding of this study determines that the direction of the qualitative data collection and analysis is to confirm and refine this distribution by exploring the relativistic thinking and commitment to relativistic thinking in particular and also to investigate the factors associated with these two styles of thinking.

Second, the survey results facilitate the sampling process. Using the average scores of each subscale (*Dualism, Multiplicity, Relativism,* and *Commitment*) gathered in the survey results, respondents were then categorized into different groups according to their prominent styles of epistemological thinking. Using this result, the purposeful sampling process in the qualitative research can be performed among the respondents in the groups of research interest.

Third, the results can help inform the interview process. Along with the scores in each subscale, the survey has also provided demographic information for each respondent, such as the respondent's academic progress, work experiences, prior undergraduate institutions, and so on. This variety of information can also help inform the interview process.

References

Baxter Magolda, M. B. (1987). Comparing open-ended interviews and standardized measures of intellectual development. *Journal of College Student Personnel, 28*, 443–448.

Cox, M. F., London, J. S., Ahn, B., Zhu, J., Torres-Ayala, A. T., Frazier, S., & Cekic, O. (2011). Attributes of Success for Engineering Ph.D.s: Perspectives from Academia and Industry. In *2011 Proceedings of the American Society for Engineering Education, Vancouver, BC, Canada.*

Culver, R. S., & Hackos, J. T. (1982). Perry's model of intellectual development. *Engineering Education, 72*, 221–226.

Golde, C. M., & Walker, G. E., (2006). *Envisioning the future of doctoral education: Preparing stewards of the discipline.* Carnegie Essays on the Doctorate, Jossey-Bass-Carnegie Foundation for the Advancement of Teaching, San Francisco, CA.

Liu, Y. (2005) *Conceptual ideas for the classifications of macro-regional development zones: A foundation for Chinese comprehensive economic districts and a complete four-level regional economic system.* Retrieved from: http://em.scnu.edu.cn/rce/macro_finance/macro_finance_62. htm (In Chinese).

Marcus, B., Bosnjak, M., Lindner, S., Pilischenko, S., & Schutz, A. (2007). Compensating for low topic interest and long surveys: A field experiment on nonresponse in web surveys. *Social Science Computer Review, 25*(3), 372.

MYCOS, My China Occupational Skills. (2010). 2010 Report on Chinese University Students' Employment.

Pavelich, M. J., & Moore, W. S. (1996). Measuring the effect of experiential education using the Perry model. *Journal of Engineering Education, 85*(4), 287–292.

Perry, W. G. (1970). *Forms of intellectual and ethical development in the college years: A scheme.* New York: Holt, Rinehart and Winston.

Survey of Earned Doctorates. (2012). Retrieved from http://www.norc.org/Research/Projects/ Pages/survey-of-earned-doctorates-%28sed%29.aspx

Wang, W. (2009). Chinese international students' cross-cultural adjustment in the U.S.: The roles of acculturation strategies, self-construals, perceived cultural distance, and english self-confidence. Retrieved from ProQuest, The University of Texas at Austin.

Wise, J., Lee, S. H., Litzinger, T. A., Marra, R. M., & Palmer, B. (2004). A report on a four-year longitudinal study of intellectual development of engineering undergraduates. *Journal of Adult Development, 11*, 103–110.

Wood, P., & Kardash, C. (2002). Critical elements in the design and analysis of studies of epistemology. In B. K. Hofer & P. R. Pintrich (Eds.), *Personal epistemology: The psychology of beliefs about knowledge and knowing* (pp. 231–260). Mahwah, NJ: Erlbaum.

Zhang, L. F. (1995). *The construction of a Chinese language cognitive development inventory and its use in a cross-cultural study of the Perry Scheme*. Retrieved from ProQuest, The University of Iowa.

Zhang, L. F. (1999). A comparison of U.S. and Chinese university students' cognitive development: The cross-cultural applicability of Perry's theory. *The Journal of Psychology, 133*(4), 425–439.

Zhang, L. F. (2000). Are thinking styles and personality types related? *Educational Psychology, 20* (3), 271–283.

Zhang, L. F. (2002). Thinking styles and cognitive development. *The Journal of Genetic Psychology, 163*(2), 179–195.

Zhang, L. F. (2004). The Perry scheme: Across cultures, across approaches to the study of human psychology. *Journal of Adult Development, 11*(2), 123–138.

Zhang, L. F., & Hood, A. B. (1998). Cognitive development of students in China and USA: opposite directions? *Psychological Reports, 82*, 1251–1263.

Zhang, L. F., & Watkins, D. (2001). Cognitive development and student approaches to learning: An investigation of Perry's theory with Chinese and U.S. university students. *Higher Education, 41*, 239–261.

Chapter 8
Factors Related to Epistemological Development

In addition to an overall picture of the Chinese engineering students' epistemological developmental profile, the possible relations between students' epistemological development and some other factors, such as their academic progress, have also been explored. Specifically, the survey results have also been analyzed with respect to the variables in the demographic survey to explore the potential relationships that exist between the epistemological development profiles and key variables, such as their academic progress, gender, and so on. This chapter illustrates the relationship between students' epistemological and demographic factors.

8.1 Academic Progress

The US doctoral education, especially engineering doctoral education, highlights its training of independent researchers with different abilities. These abilities include creating original research ideas, absorbing and discerning the knowledge accumulated from both past history and the latest technological advances, communicating research findings and/or technological advances to both the scholarly community and a larger audience in both oral and written manners, and adopting and applying latest research findings and translating the findings into functional products or procedures (Golde and Walker 2006). These trainings do not offer a promise, but they still strongly imply the potential benefits for the students to develop relativistic thinking and move toward higher levels in their epistemological development. Here, the possible correlation between the students' epistemological developmental stages and their academic progress is explored in a quantitative manner.

© Springer Science+Business Media Singapore and Higher Education Press 2017
J. Zhu, *Understanding Chinese Engineering Doctoral Students in U.S. Institutions,*
East-West Crosscurrents in Higher Education, DOI 10.1007/978-981-10-1136-8_8

8.1.1 Overall Picture

A breakdown of the number of students in each group across their academic progress is shown here in Table 8.1. A visual representation of the percentage of students within each group across their academic progress is illustrated in Fig. 8.1.

In the course work stage, there are 20 % in the group of *Dualism, Multiplicity,* and *Multiplicity-Relativism,* while 73 % are in *Relativism, Relativism-Commitment,* and *Commitment.* At the stage when they have passed their qualifying and/or preliminary examinations or other equivalent examinations, there are 12 % in *Dualism, Multiplicity,* and *Multiplicity-Relativism* and 84 % in *Relativism, Relativism-Commitment,* and *Commitment.* At the dissertation stage, there are 12 % in *Dualism, Multiplicity,* and *Multiplicity-Relativism* and 80 % in *Relativism, Relativism-Commitment,* and *Commitment.* The trend seems to suggest that for those students who have not already reached relativistic thinking when they entered the program, they could still develop toward a higher level of thinking. However, because many students have already scored high in the higher level of thinking

Table 8.1 A breakdown of the number of students in each group across their academic progress

Academic progress	Groups							Total
	D	M	M-R	R	R-C	C	Other	
1	4	5	3	21	11	11	4 (2 D-M, 1 D-R,1 M-R-C)	59
2	2	2	2	33	5	7	2 (1 D-R, 1 D-C)	53
3	1	2	0	13	3	4	2(1 D-R, 1 M-R-C)	25
Total	7	9	5	67	19	22	8	137

Note 1, course work stage; 2, passed qualifying and/or preliminary examinations or other similar milestone examinations; 3, dissertation stage
Abbreviations: D, *Dualism*; M, *Multiplicity*; R, *Relativism*; C, *Commitment*

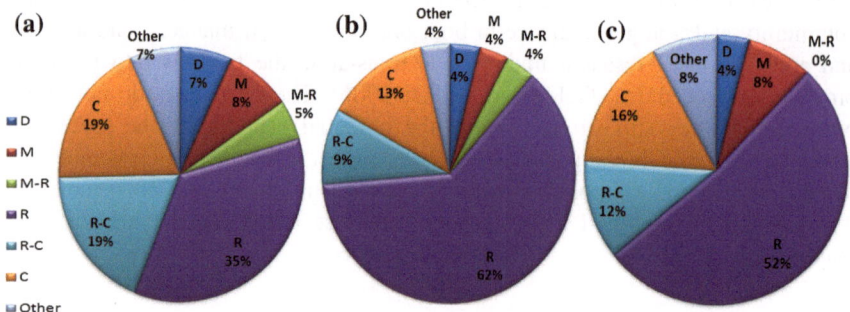

Fig. 8.1 Percentage of students in different groups across their academic progress: **a** course work stage; **b** passed qualifying and/or preliminary examinations or other similar milestone examinations; **c** dissertation stage. Abbreviations: D, *Dualism*; M, *Multiplicity*; R, *Relativism*; C, *Commitment*

stages (*Relativism* and *Commitment*) when they are at the course work stage of the doctoral program, it becomes increasingly difficult to reach a firm conclusion on the development trend without additional statistical testing.

8.1.2 ANOVA

One-way ANOVA procedures were used to test for significant difference in the four subscales (*Dualism*, *Multiplicity*, *Relativism* and *Commitment*) (dependent variables) with regards to the students' academic progress. It seems that a significant difference exists for the *Dualism* subscale between the scores of the students at the course work stage and those of the students who passed their qualifying and/or preliminary examinations or other equivalent examinations (p value $= 0.030 < 0.05$, $F = 3.595$, observed power $= 0.657$, Table 8.2). No significant difference can be observed for the other two group comparisons. Mean scores on the *Dualism* subscale for students with different academic progresses are shown in Table 8.3.

Pairwise comparisons showed that compared to the students at the course work stage, the students who have passed qualifying and/or preliminary examinations or other equivalent examinations score lower in the *Dualism* subscale (Table 8.4). To summarize, the ANOVA results of students' scores in each subscale by academic progress illustrate some significant differences within the scores of the *Dualism* subscale. Therefore, it is reasonable to say that the Chinese engineering students who undergo the doctoral program for a few semesters showed a lower level of dualistic thinking when compared to their in-coming peers.

Table 8.2 One-way ANOVA: *dualism* (DV) by differing levels of academic progress (IV)

Source	Type III sum of squares	df	Mean square	F	Sig.	Noncent. parameter	Observed power
Academic progress	1.554	2	0.777	3.595	0.030	7.190	0.657

Table 8.3 Mean score in *dualism* subscale for different academic progresses

Academic progress	Mean	Std. error	95 % confidence interval	
			Lower bound	Upper bound
1	2.909	0.061	2.790	3.029
2	2.675	0.064	2.549	2.802
3	2.762	0.093	2.578	2.946

Note 1, course work stage; 2, passed qualifying examinations and/or preliminary examinations or other similar milestone examinations; 3, dissertation stage

Table 8.4 Pairwise comparison for the *dualism* subscale for different academic progresses

(I) Academic progress	(J) Academic progress	Mean difference (I-J)	Std. error	Sig.	95 % confidence interval	
					Lower bound	Upper bound
1	2	0.234	0.088	0.032	0.016	0.452

Note 1, course work stage; 2, passed qualifying and/or preliminary examinations or other similar milestone examinations

To summarize, the results based upon the quantitative data suggest that students who are in a later stage of the academic program seem to exhibit lower levels of dualistic thinking. I would speculate that the new students whose prominent thinking styles reflect a dualistic or multiplistic manner of thinking are much more likely to undergo changes during their first few semesters at school. However, because a large number of students who are in their early stage of development have shown a high level of relativistic thinking, an in-depth qualitative study is needed to explore the validity of this preliminary speculation and to truly understand the impact of academic progress on the students' epistemological developmental progress.

8.2   Places of Origin

In the demographic session, the students answered the question, "Which province (or region) of China are you from?" According to a classification modified from the Chinese government's office report (Liu 2005), the region of Mainland China can be roughly classified into ten major areas based on the differences in population density, urban spatial distribution, economic development, and other related factors. These ten areas are:

1. *Northeast district (Liaoning, Jilin, and Heilongjiang)*
2. *North coastal district (Beijing, Tianjing, Hebei, and Shandong)*
3. *East coastal district (Shanghai, Jiangsu, and Zhejiang)*
4. *Southeast coastal district (Guangdong, Fujian, and Hainan)*
5. *Upper and middle reaches of the Yellow River (Shanxi, Gansu, Ningxia, Shaanxi, and Henan)*
6. *Upper and middle reaches of the Yangtze River (Sichuan, Chongqing, Hubei, Hunan, Anhui, and Jiangxi)*
7. *Upper and middle reaches of the Pearl River (Yunnan, Guizhou, and Guangxi)*
8. *Inner Mongolia district*
9. *Xinjiang district*
10. *Qingzang Altiplano district (Qinghai and Xizang)*

According to this classification, the students' answers were then assigned to their corresponding regions of Mainland China. With "Place of Origin" as independent variable, the one-way ANOVA results indicate significant differences for the Relativism subscale by Place of Origin (Table 8.5). Mean scores in the Relativism subscale for students from different regions are shown in Table 8.6.

Pairwise comparisons show that several regions' means on the Relativism subscale have significant differences (Table 8.7). Students from the Region 6 Upper and Middle Yangtze River district (Sichuan, Chongqing, Hubei, Hunan, Anhui, and Jiangxi) have higher Relativism scores than those from Regions 1 and 2.

This observation suggests the differences in students' epistemological development are related to additional factors associated with the region in which they grew up before they entered the graduate school. The differences in their geographic regions in which they grew up can potentially be associated with other factors like economics and infrastructures. It is reasonable to conclude that differences in students' epistemological development may also be related to factors existent before their doctoral education.

8.3 Prior Master's Education

Prior graduate training, specifically having obtained a master's degree, could potentially have a positive effect on the students' epistemological development. The emphasis on graduate training lies in becoming an independent researcher and

Table 8.5 Difference in *Relativism* subscale for different places of origin

Source	Type III sum of squares	df	Mean square	F	Sig.	Noncent. parameter	Observed power
Place of origin	2.115	7	0.302	3.377	0.003	23.639	0.955

Note $n = 129$, 8 students' information on their places of origin were not available

Table 8.6 Mean score on the *relativism* subscale for different regions

Region	Mean	Std. error	95 % confidence interval	
			Lower bound	Upper bound
6	3.843	0.050	3.744	3.942
5	3.650	0.077	3.497	3.803
3	3.649	0.053	3.544	3.753
7	3.640	0.122	3.398	3.882
4	3.615	0.095	3.428	3.802
2	3.518	0.070	3.378	3.657
1	3.455	0.090	3.277	3.634
8	3.420	0.299	2.828	4.012

Note No participants are from the last two of the ten areas

Table 8.7 Multiple comparisons: difference in *relativism* subscale for different regions

(I) Region	(J) Region	Mean difference (I-J)	Std. error	Sig.	95 % confidence interval	
					Lower bound	Upper bound
6	1	0.388	0.103	0.034	0.0154	0.760
6	2	0.325	0.086	0.034	0.0134	0.637

Table 8.8 Difference in *multiplicity* (DV) by whether students have prior masters' degrees or not (IV)

Source	Type III sum of squares	df	Mean square	F	Sig.	Noncent. parameter	Observed power
Master's degree	1.303	1	1.303	5.647	0.019	5.647	0.654

Note n = 114 (23 students' information is not available here)

Table 8.9 Average *multiplicity* subscale scores for prior master's degree

With a MS	Mean	Std. error	95 % confidence interval	
			Lower bound	Upper bound
Yes	3.329	0.066	3.198	3.460
No	3.115	0.062	2.993	3.237

developing the skills of innovative thinking and achievement. Although not all of these emphases are listed within the specific educational goals of a master's student, they can still be partially instilled with students who went through the training.

One-way ANOVA procedures were used to test for significant difference in the four subscales (*Dualism, Multiplicity, Relativism,* and *Commitment*) (dependent variables) with regards to whether students have a prior master's degree or not. It seems that significant difference exists for the *Multiplicity* subscale ($0.019 < 0.05$, Table 8.8). Students with a prior master's degree score higher in the *Multiplicity* subscale (Table 8.9).

8.4 Enrolled Universities

This study includes student data from five different universities as the sample. With one-way ANOVA testing of the four subscales for the differing universities, the *Dualism* subscale scores for students from these different universities demonstrate a statistically significant level of difference ($0.021 < 0.050$, see Table 8.10). Mean scores for the *Dualism* subscale for students from different current universities is shown in Table 8.11. Multiple comparisons indicate that students from two

Table 8.10 Differences in *dualism* (DV) by different current universities (IV)

Source	Type III sum of squares	df	Mean square	F	Sig.	Noncent. parameter	Observed power
Current university	2.455	4	0.614	2.998	0.021	11.992	0.784

Note n = 121 (16 students' information is not available here)

Table 8.11 Mean scores for the *dualism* subscale for each different current university

Current university	Mean	Std. error	95 % confidence interval	
			Lower bound	Upper bound
University D	2.936	0.171	2.597	3.274
University A	2.933	0.073	2.788	3.078
University C	2.807	0.096	2.616	2.998
University E	2.798	0.096	2.607	2.989
University B	2.572	0.080	2.413	2.730

Table 8.12 Pairwise comparison: differences in the *dualism* subscale for each different current university

(I) Current university	(J) Current university	Mean difference (I-J)	Std. error	Sig.	95 % confidence interval	
					Lower bound	Upper bound
University A	University B	0.361	0.109	0.031	0.021	0.701

universities out of five show statistically significant differences in the *Dualism* scores (see Table 8.12).

The difference in the scores for the *Dualism* subscale for students from the differing universities suggests the need to include students from different programs in the qualitative data collection because of the potential variations of students from differing universities. It also signifies the importance for future studies to consider the variations that are derived from the specific differences within the students' academic departments or universities. Although no clear conclusion or speculations can be drawn from this single observation, it offers another perspective from which to consider the potential variations in the student sample.

8.5   Discussion

The series of one-way ANOVA tests have shown several factors that have exhibited statistically significant differences in some of the subscales. These factors include the students' academic progress, their places of origin, their prior achievement of a master's degree, and their current universities. These observations suggest that such various factors may possibly be related to the students' epistemological development profiles.

A causal relationship is not assumed between these factors and the students' epistemological development, but the identification of these factors helps one to consider the aspects that could be potentially related with the epistemological development of the Chinese engineering doctoral students.

Concerning the relationship between academic progress and the students' epistemological development, students who were in their earlier stage of study (course work) seemed to reflect an increased propensity toward dualistic thinking than did those who were further along in their academic progress (having passed milestone examinations and/or at the dissertation stage). It seems that in the first few semesters, students start to gain an understanding about being an independent researcher and strive toward achieving this role. As doctoral students, the students are trained in terms of their analytical thinking skills, innovative thinking skills, abilities to conduct rigorous research, abilities to communicate their scientific findings to a scholarly community, and growing ability to translate theory into application, and so on (Golde and Walker 2006). Many of these training skills could potentially promote their thinking to grow along Perry's scale toward a more relativistic style of thinking, which emphasizes a reasoning process in which different factors, evidence, and constraints are critically evaluated. Therefore, going through this type of doctoral educational program can potentially promote a student's epistemological development toward a higher level of thinking in Perry's theory.

As to the relationship between these students' epistemological development and their places of origin, it appears that students from some regions of China reflect an increased tendency toward relativistic thinking than is found in other regions. This finding implies that complicated causal relationships or other correlations may be present to account for said students' epistemological development and the factors that are associated with their places of origin. The differences in terms of the geographical locations can potentially be associated with varied economic development levels, social and educational infrastructures and resources, cultural values, and other factors that can be possibly related with one's epistemic thinking (Weinstock 2010). Therefore, on one hand, this observation may indicate the necessity to include and control for factors that are related to the regions from which the students originate in an epistemological study; on the other hand, this observation alone cannot secure any clear conclusion on the relationships between a region and its associated factors and the students' epistemological development.

With regards to the relationship between students' reception of prior master's degrees and their epistemological thinking, the results do indicate that students with a prior master's degree reflect more multiplistic thinking than do those who without one. A multiplistic style of thinking emphasizes the legitimacy of differing viewpoints. This type of thinking tends to permit a higher tolerance for ambiguity than does dualistic thinking. For someone to have a higher tendency toward multiplistic thinking, it often becomes essential that he/she is first exposed to a diversity of views. A graduate school can potentially provide this diversity of views by engaging the graduate students in course-related discussions/activities or other types of collaborations in research.

The last factor that was identified by the one-way ANOVA tests was these students' current universities of attendance. It appears that students from differing universities seem to have a statistically significant difference in their levels of dualistic thinking. This finding suggests that additional complicated factors associated with the students' current universities can be potentially related these students' abilities for epistemological thought. Indeed, a number of factors that are associated with the students' current universities, such as available academic resources, the departmental culture, and other elements, will require further investigation to understand their impact on the students' level of epistemological development.

8.6 Conclusion

To summarize, the results from the quantitative data indicate some factors that are potentially related with students' epistemological development, including the students' academic progress, places of origin, current universities, and prior achievement of a master's degree. These different factors provide some insight into the potential factors that can facilitate the students' epistemological development. To better understand the specific factors that are associated with the development toward relativistic thinking among Chinese doctoral students, further investigation and in-depth qualitative research are required. Some demographics, such as places of origin and current universities, may also require further explorations on the policy implications.

In addition to the implications to the studies in epistemological development, findings from current studies can inform and facilitate the qualitative data collection and analysis. The survey results can assist in the design of the interview protocol. The factors that were identified to be related with the students' epistemological development in the survey process could potentially be areas for further explorations within a semi-structured interview protocol.

Moreover, according to the average scores of the subscales, some other factors did not show statistically significant differences, for example, work experiences. It is possible, however, that students who have had past work experiences could actually show higher level of relativistic thinking. These areas can be further explored in the semi-structured interview.

References

Golde, C. M., & Walker, G. E. (2006). *Envisioning the future of doctoral education: Preparing stewards of the discipline.* Carnegie Essays on the Doctorate, Jossey-Bass-Carnegie Foundation for the Advancement of Teaching, San Francisco, CA.

Liu, Y. (2005). Conceptual ideas for the classifications of macro-regional development zones: A foundation for chinese comprehensive economic districts and a complete four-level regional economic system. Retrieved from: http://em.scnu.edu.cn/rce/macro_finance/macro_finance_62.htm (In Chinese)

Weinstock, M. (2010). Epistemological development of Bedouins and Jews in Israel: Implications for self-authorship, In M. B., Baxter Magolda, E. G., Creamer & P. S., Meszaros (Eds.), *Development and assessment of self-authorship: Exploring the concept across cultures.* Sterling, VA: Stylus.

Part IV
A Qualitative Exploration of Chinese Engineering Doctoral Students' Epistemological Thinking Styles

Chapter 9
Stories of Chinese Engineering Doctoral Students' Epistemological Thinking Styles

Building upon the results from the quantitative data analysis, the purpose of this study is to both confirm and refine what is understood as the epistemological thinking of Chinese engineering doctoral students among the survey respondents. Another goal is to explore the factors that are related to the trends in the epistemological developmental status among the Chinese engineering doctoral students. In this chapter, I describe the details of qualitative data collection, data analysis, and present stories of Chinese engineering students' epistemological thinking styles. The results obtained from this study will refine the results from the quantitative data and provide an in-depth picture of Chinese engineering doctoral students' epistemological development through the use of qualitative data and analysis.

9.1 Background

As was found from the quantitative data, the most prominent style of thinking for nearly half of the Chinese engineering doctoral students is relativistic thinking. Most of the students' prominent thinking styles fell into the higher level of thinking (i.e., *Relativism* or *Commitment*) according to Perry's theory (nearly 80 %), while only around 15 % of the students' prominent thinking styles fell into the lower levels of the Perry's theory (i.e., *Dualism* or *Multiplicity*). The development toward relativistic thinking is possibly related to the students' current exposure to academic experiences within graduate school, prior experiences when working to achieve a master's degree, and other factors like their places of origin or their current universities.

Again the research questions ask:

1. *What are the epistemological development profiles of Chinese engineering doctoral students framed within the context of the modified Perry's theory on epistemological development?*

© Springer Science+Business Media Singapore and Higher Education Press 2017
J. Zhu, *Understanding Chinese Engineering Doctoral Students in U.S. Institutions,*
East-West Crosscurrents in Higher Education, DOI 10.1007/978-981-10-1136-8_9

2. *Based on research results of the research question 1, what are the possible factors that are related to these profiles?*

The design of this research work is explanatory in nature (Creswell 2008). Therefore, the attempt to collect and analyze qualitative data is highly dependent on the results obtained from Chaps. 7 and 8. Qualitative data collection is used to explain key results from quantitative data analysis or to specifically refine the results by exploring typical cases and some cases of outliers.

The goal of this chapter focuses on confirming and refining the results from the quantitative data analysis. That is, results from qualitative data collection and analysis are aimed to confirm the epistemological developmental stages of sampled participants and to refine the different styles of epistemological thinking by operationalizing the demonstration of each epistemological thinking style.

Few past qualitative studies have focused on understanding the epistemological developmental status of the Chinese engineering doctoral students. Nonetheless, the past literature available regarding Chinese engineering doctoral students' adjustment issues has shown some implications with regards to the students' epistemological development status. Jiang (2010) conducted one such study that is particularly relevant here.

Jiang (2010) conducted in-depth interviews with ten Chinese engineering doctoral students in a Midwestern university regarding these students' perceptions of their cross-culture adaption experiences. Some of her findings were regarding areas in which students found difficulty in their cross-cultural adaption, such as lack of English proficiency, which is similar to the findings of former researchers on the topic of cross-cultural adaption (Ye 1992; Wang 2009). More importantly, Jiang's study also provided some indirect evidence that is particularly relevant to the understanding of these students' epistemological development.

First, Jiang described the participants' appreciation of the benefits related to learning technical writing when most of these students had found technical writing to be challenging early on in their doctoral studies. Some of these students had suggested that the advisors had played an important role in this process. Most of them mentioned reading literature as a common strategy. These findings regarding the Chinese engineering doctoral students' learning experiences suggest some meta-cognition processes of adjusting learning strategies and potentially some epistemological level of thinking and processing of knowledge and information.

Moreover, Jiang presented some evidence regarding the excitement of some Chinese students' development as independent researchers in their experiences here in US engineering graduate programs. Some students contrasted their current experiences with their previous graduate studies in China. Some students commented on the lack of freedom in their research assignments while in China and their surprise regarding the amount of freedom they found in the US. They stated that while they had received some guidance from their advisors, it had remained up to them to discover solutions in their research. This finding suggested that these students were learning to think independently, an action that is also an important part of the epistemological developmental process as described in Perry's theory.

Informed from these prior findings and building upon the results gathered from the quantitative data, this study has adopted a preferable methodological framework, phenomenology, in understanding students' epistemological development. It has been used and supported by a number of researchers, including Perry (Perry 1970; Belenky et al. 1986; Baxter Magolda 1992), in the study of epistemological development. The use of phenomenology allows researchers to understand students' perceptions of their own experiences in their own terms, a process that fits exactly with the purpose of developing an understanding of the students' epistemological development through their own perspectives.

9.2 Methods

9.2.1 Data Collection

Sampling

The method of purposeful sampling was used in this study. Recruiting emails were sent to all of the survey participants who satisfied the following two criteria: (1) according to the quantitative analysis results, they fit into one of the following groups: *Dualism*, *Multiplicity*, *Relativism*, *Relativism-Commitment*, or *Commitment*; (2) they indicated their interest of being interviewed in the initial survey. Nineteen interviewees were collected after the recruiting efforts. A list of interviewees, their pseudonyms, their average scores for the survey items for each epistemological stage and the Knowledge Construction and Modification (KCM) subscale of the Epistemological Belief Survey, and other demographic information is shown in Table 9.1.

Out of all 19 participants, 6 are from the Department of Electronic, Computer, and Electrical Engineering; 4 are in Mechanical Engineering; 4 are in Biomedical Engineering; 1 is in Civil Engineering; 1 is in Industrial Engineering; 1 is in Materials Science Engineering; 2 are from other engineering majors.

Out of all 19 participants, 16 are single, 2 are married, and 2 are in committed relationships.

Out of all 19 participants, 15 are from urban areas, 3 are from suburban areas, and 1 is from a rural area.

Out of all 19 participants, 12 are from University A, 4 are from University E, and 3 are from University B.

Out of all 19 participants, 8 indicated that their religion is Atheism; 8 indicated that their religion is Christianity; 3 did not provide this information.

For the highest educational level of the participant's father, one indicated this level to be less than secondary school; five indicated it as secondary/high school; one indicated it as some college; seven indicated it as a bachelor's degree; and five indicated it as a master's degree.

Table 9.1 Interview participants

Grouping	Pseudonym	Gender	Age	Academic progress	With a master's degree (Y/N)	Work experiences	D	M	R	C	KCM
D	David	M	25–30	2	N	0	4.2	4.11	2.5	2.14	2.73
M	Mike	M	25–30	3	Y	0	2.35	3.67	3.58	3.14	4
M	Mary	F	22–25	1	N	0	2.7	3.56	2.83	2.86	3.36
R	Rick	M	22–25	2	N	0	2.9	3.56	3.92	3.57	4.64
R	Robert	M	22–25	1	Y	0	2.7	2.78	3.92	3.71	4.09
R	Ron	M	22–25	2	N	0–1 year	3.2	3.67	3.83	3.57	3.64
R	Rena	F	25–30	3	Y	0–1 year	2.45	3.11	3.83	3.57	3.73
R	Ryan	M	25–30	1	N	0	2.6	2.33	3.83	3	3.82
R	Ruby	F	25–30	1	Y	0–1 year	2.7	3.78	3.75	2.86	4
R	Rose	F	22–25	1	Y	0	3.3	3.44	3.75	3.29	4.09
R	Rebekah	F	25–30	2	N	0	3	3	3.5	3.29	3.09
R	Ray	M	25–30	3	N	0	2.45	2.67	3.5	3.14	3.64
R-C	Ken	M	22–25	1	Y	0	2.85	3.44	3.92	4.29	4.73
R-C	Kirk	M	25–30	3	Y	0	2.5	2.89	3.58	4	4.27
R-C	Kevin	M	30–35	3	Y	1–3 years	2.55	3.11	3.92	3.86	4
C	Cameron	M	22–25	3	N	0–1 year	3.4	3.89	3.75	4.57	4.55
C	Cody	M	25–30	2	Y	0	2.5	2.78	3.42	4.14	4.18
C	Charles	M	<22	1	N	0–1 year	2.05	2.33	3.58	4	4.09
C	Charlie	M	25–30	3	Y	0	2.45	3	3.75	3.86	4.36

Note

Academic progress: *1* course work; *2* passed qualifying and/or preliminary examinations or other similar milestone examinations; *3* dissertation stage

Abbreviations

D *Dualism*; M *Multiplicity*; R *Relativism*; C *Commitment*; KCM *Knowledge Construction and Modification subscale*

For the highest educational level of the participant's mother, one indicated this level to be less than secondary school; seven indicated it as secondary/high school; two indicated it as some college; eight indicated it as a bachelor's degree; one indicated it as a master's degree;

Interview

The interviews were conducted approximately one and a half months after the initial quantitative survey. Each interview lasted around an hour. All interviews were conducted primarily in the Chinese language, with the exception of one interview, which was conducted in English. For the single interview that was conducted in English, the interviewee had indicated that he was more comfortable speaking English. For all of the other interviews that were conducted primarily in Chinese, both the interviewee and the interviewer frequently inserted English words throughout the interviews to better convey his/her ideas.

Interview Protocol

Conducting a qualitative interview is a preferable manner by which to study each individual's epistemological thinking; the qualitative interview process has been adopted by most researchers within this research field, most notably during the early stage when the epistemological developmental theory was being developed and extended in different ways (Perry 1970; Belenky et al. 1986; Baxter Magolda 1992). A phenomenological framework was adopted in these studies because the essence of a phenomenological method lies in understanding the participants' lived experiences in their own terms (Patton 2002). The phenomenological methodological framework is useful for guiding the design of the interview protocol in this study because it provides rich in-depth details about the students' lived experiences concerning their epistemological understanding in their own terms. A complete interview protocol can be found in Appendix D in both Chinese and English.

The interview protocol here has been modified mainly from Baxter Magolda's interview protocol (1992). Her protocol serves as an appropriate main reference because it retains the advantages in Perry's original design of the unstructured interview. In addition, it adds some structure to the protocol by dividing the content area into six sub-areas (i.e., role of learners, role of instructors, role of peers, perception of evaluation of their work, nature of knowledge, and educational decision making). Within each sub-area, she retained the unstructured interview procedure, where interview questions were asked to introduce the topic but not frame the response. The primary advantage of this strategy is the fact that it introduces few presuppositions or potential biases; instead, a clear boundary is established for the topic area. By using this method of interview, six initial topics were included in the protocol (see Appendix D). Perry's original interview (i.e., "What stands out in your learning experience?") was also included as the opening question with the intent to solicit an open response without introducing a topic. With these questions, the goal of the interview is designed to confirm and refine the epistemological developmental stages of participants.

Second, the second research question in this work asks *"What are the possible factors that are related to these profiles?"* To get to the essence of possible factors that are related to the specific epistemological thinking stages, a "how" question was added following each of the initial questions employed in Baxter Magolda's six areas. For example, for the topic "Role of Instructors," the first question is, "As you think about your instructors, professors, advisor(s), what role do you think they have played that made you learn effectively?" The follow-up question for this topic reads, "How did you realize that these roles/functions are useful for your learning?"

Third, because the different stages of thinking as described in Perry's theory often involve differing responses when a person encounters the various viewpoints, an explicit question was added toward the end of the interview protocol to solicit responses to differing viewpoints. This method has also been used in one of Baxter Magolda's interview protocols (1992).

Fourth, according to the survey results, the students' experiences prior to their years in college are likely to be related to their epistemological development. Therefore, an additional question was added to explore the potential factors that are related to their epistemological development. This question reads, "What kinds of factors influenced your learning when you grew up?"

Last but not least, considering the fact that a major portion of the Chinese engineering doctoral students fell within the later stages of the Perry's scale, that is, the *Relativism* and/or *Commitment* groups, it is important to explore their commitment to relativistic thinking in areas other than education. This is because, for the higher levels of thinking according to Perry's scale, i.e., Positions 6–9 (see definitions in 2.2 Perry's Theory), the differences of these positions lie in the extent to which relativistic thinking is applied to education and areas other than education. Therefore, another two interview questions have been added to explore their decision-making processes in education and areas other than education (see Appendix F).

9.2.2   Data Analysis

Transcription

All of the interviews were transcribed by the author. The recordings and transcriptions were reviewed repeatedly to develop an initial understanding of each transcript.

Coding Procedure

This study was designed in the context of Perry's theory. The theoretical framework should also guide the data analysis process. Moreover, the research questions and prior findings based upon the quantitative data should be taken into considerations before the data analysis. Specifically, two categories, including five mega-codes, were defined as a priori codes (Table 9.2).

These meta-codes were defined because of the following reasons and considerations:

Table 9.2 Five codes were pre-defined as a priori codes

Codes (e.g., first level code–second level code)	
Demonstrations	Factors
Dualistic thinking—demonstration	
Multiplistic thinking—demonstration	
Relativistic thinking—demonstration	Relativistic thinking—factor
Commitment—demonstration	

First, this study was designed in the context of the Perry's theory. According to Merriam (2006), the theoretical framework not only determines the research questions asked and the data set to be collected to address the research question, but, more importantly, it should also guide the data analysis process; specifically, it guides the coding process for this current study. Four major groups of codes can be used to guide the coding about thinking styles, which are dualistic thinking, multiplistic thinking, relativistic thinking, and commitment (in relativistic thinking). These codes should serve as the first level of grouping of codes in the coding process.

Second, in addressing the two research questions of this study, the qualitative research plays two major roles or has the following two goals:

(1) To confirm the epistemological developmental stages of participants and refine the epistemological thinking by operationalizing the demonstration of each epistemological thinking style and
(2) To explore those factors that are related to each particular epistemological developmental stage.

The goals and framework in this qualitative research should guide the data analysis process before any open-ended coding comes in. Therefore, two categories of codes can be used to address the two research goals: *Demonstrations* and *Factors*. Through the use of the word *"Demonstrations,"* the research purpose is to then identify activities, behavioral patterns, or any direct expression of thinking process that reflects the thinking of the first level codes. By using the word "factors," the goal may refer to any factors (people, events, location, etc.) that are potentially related to the development of the particular manner of thinking.

Third, the results based upon the quantitative data suggest that a major portion of the students fell into the group of *Relativism*. Therefore, to understand the factors that are associated with the relativistic thinking has become the major purpose of this study. The results of related factors are presented in Chap. 10.

In sum, these considerations determine that the coding process is a guided procedure instead of a completely open-coding process. To achieve the goals of this study, it relies heavily on the exploration of these five a priori codes.

Within the next level of coding, open coding was used to operationalize each thinking style, i.e., to identify the actual demonstrations for each thinking style. Open coding was also used to explore the specific factors that are related to relativistic thinking. These five predefined codes were used across all of the transcripts wherever the demonstration of that particular code was found within the transcript.

Codebook Development

The importance of a structured coding process has been pointed out by multiple qualitative researchers (Miles and Huberman 1994; MacQueen et al. 1998). A structured codebook can help researchers in their coding process by offering a relatively stable frame (MacQueen et al. 1998).

To develop a codebook, the initial coding was done for four transcripts with one transcript from each of the following four groups: *Dualism, Multiplicity, Relativism,* and *Commitment.* The choice of the transcripts for the initial coding was based on the initial understanding of each transcript. Transcripts that were rich in conversation content were chosen for initial coding. After the coding of the four transcripts, an initial codebook was developed.

Inter-coder Reliability

Inter-coder reliability is a measure used to assess the degree to which codes of transcripts assigned by different coders agree with each other when these coders used the same codebook for their coding (Hruschka et al. 2004).

After finishing the coding of the first four transcripts, there were 66 different codes in the initial codebook. Quotes were chosen arbitrarily by using a random half of the codes' corresponding quotes. This procedure resulted in 26 different quotes derived from all four transcripts (some codes link to the same quote).

One graduate student researcher who was experienced with qualitative research but not familiar with this research was invited to code these quotes using the initial codebook independently. An auditing procedure was conducted to discuss the results of the coding. Codes are modified and/or refined during and after the auditing process.

The inter-coder reliability was calculated to compare the codes by the lead coder and the external coder to the codes agreed upon after the discussion between them. The inter-coder reliability was calculated for the first, second, and third level codes, respectively, resulting in agreement of 100 % for the lead coder and 92.8 % for the external coder (first level codes), 96.8 % for the lead coder and 100 % for the external coder (second level codes), and 75.9 % for the lead coder and 67.2 % for the external coder (third level codes).

It is suggested across the literature that calculating simple percentage agreement can potentially overestimate the degree of inter-coder reliability without the consideration of chance. Cohen's kappa is a measure introduced to take the presence of chance in the coding process into consideration (Cohen 1960; Gwet 2012). Here, Cohen's kappa was calculated to compare the codes by the lead coder and the external coder to the codes agreed upon after the discussion between them. Cohen's kappa was calculated for the first, second, and third level codes, respectively, resulting in a kappa value of 1 for the lead coder and 0.89 for the external coder (first level codes), 0.91 for the lead coder and 1 for the external coder (second level codes), and 0.75 for the lead coder and 0.66 for the external coder (third level codes). Based on the results of percent agreement and Cohen's kappa, the inter-coder reliability fell within the range of good to excellent (Landis and Koch 1977; Cicchetti 1994).

The code for the Relativistic Thinking-Factor was chosen to code the factors that were determined to be related to both relativistic thinking and commitment to relativistic thinking. This choice was based on the fact that a "commitment" means a commitment to the style of relativistic thinking. The demonstrations of relativistic thinking and commitment are closely intertwined in the transcript. Conceptually, it is very difficult to separate the factors that are related with relativistic thinking and the factors that are related with a commitment to relativistic thinking. Therefore, the code of Relativistic Thinking-Factor was determined to be necessary to code the factors that are related both to relativistic thinking and commitment to relativistic thinking.

The revised codebook was used to guide the open coding among all 19 transcripts. The frequency of codes and the distributions of codes among participants were summarized. Some codes were further grouped into categories to represent a theme in the analysis (Creswell 2008).

Researcher as the Instrument

In qualitative research, Merriam (2002) spoke of the qualitative researcher as "the primary instrument for data collection and data analysis;"; the researcher can be "responsive and adaptive" in the process of data collection and interpretation (p. 5). Therefore, these advantages can benefit the goal of understanding the lived experiences of the students in this study. Meanwhile, it is important to recognize the biases from the researcher and the potential impact on the research.

For this study, the researcher's past educational experiences in Chinese and US institutions allowed the researcher to better understand the participants' lived experiences. However, the researcher was aware of the potential biases that could come from the researcher's own educational experiences and actively monitored them in the processes of one-on-one interviews, data analyses, and interpretation.

9.3 Results

9.3.1 The Overall Picture

Overall, the findings based upon the qualitative data confirm the groupings of the survey results. Students who were grouped into a certain epistemological developmental group according to the survey results tended to illustrate most instances of that epistemological thinking style according to the qualitative findings, except in those cases where the students seemed to reflect two or more thinking styles at the same time. Tables 9.3, 9.4, 9.5, 9.6 and 9.7 present the counts and the percentage of each of the demonstration code from the qualitative analysis for students from different groups according to the quantitative analysis.

Table 9.3 The counts and percentages of each demonstration code in the *Dualism* group

Codes	David	
	Count	Percentage (%)
D-D (dualistic thinking—demonstration)	20	60.6
M-D (Multiplistic thinking—demonstration)	6	18.2
R-D (Relativistic thinking—demonstration)	7	21.2
C-D (commitment—demonstration)	0	0.0
Total	33	100.0

Note Abbreviations D-D, M-D, R-D, and C-D are the same as for Tables 9.4, 9.5, 9.6 and 9.7; Fig. 9.1

Table 9.4 The counts and percentages of each demonstration code in the *Multiplicity* group

Codes	Mary		Mike		Average	
	Count	Percentage (%)	Count	Percentage (%)	Count	Percentage (%)
D-D	6	26.1	4	12.9	5	17.9
M-D	8	34.8	1	3.2	4.5	16.1
R-D	8	34.8	15	48.4	11.5	41.1
C-D	1	4.3	11	35.5	6	21.4
Total	23	100.0	31	100.0	28	100.0

As shown in Table 9.3, the transcript of the participant, David, from the *Dualism* group illustrated over 60 % of his total demonstration codes as dualistic thinking. There are some instances of multiplistic thinking and relativistic thinking. No commitment to relativistic thinking was identified in his transcript.

As shown in Table 9.4, the transcript of the participant Mary from the *Multiplicity* group demonstrated approximately 35 % instances of multiplistic thinking and relativistic thinking and around 26 % instances of dualistic thinking. It should be noted that from the interview, Mary seemed to go through some transitions into relativistic thinking a week or so before the interview, which happened several weeks after she took the survey. More details will be provided in Sect. 9.3.3.

Mike, who is from *Multiplicity* group, had 48.4 % instances of relativistic thinking, 35.5 % of commitment to relativistic thinking, 12.9 % of dualistic thinking, and 3.2 % of multiplistic thinking. It should be noted that, in the results of his survey responses, he scored high on both the *Multiplicity* and *Relativism* subscales, that is, 3.67 and 3.58, respectively. Since the transition from *Multiplicity* to *Relativism* is the most significant transition within Perry's scale, for Mike to demonstrate both *Multiplicity* and *Relativism* at the same time suggests he may not be a typical example for *Multiplicity*. More details will follow in Sect. 9.3.3.

On average, the *Multiplicity* group did show a much lower representation in dualistic thinking. It also illustrated a mixed representation of different thinking styles across Perry's scale.

Table 9.5 The counts and percentages of each demonstration code in the *Relativism* group

Codes	Ray		Rebekah		Rena		Rick		Robert	
	Count	Percentage (%)	Count	Percentage (%)	Count	Percentage (%)	Count	Percentage (%)	Count	Percentage (%)
D-D	0	0.0	0	0.0	4	11.1	0	0.0	0	0.0
M-D	0	0.0	0	0.0	0	0.0	0	0.0	0	0.0
R-D	25	78.1	32	62.7	28	77.8	26	81.3	54	79.4
C-D	7	21.9	19	37.3	4	11.1	6	18.8	14	20.6
Total	32	100.00	51	100.00	36	100.00	32	100.00	68	100.00

Codes	Ron		Rose		Ryan		Ruby		Average	
	Count	Percentage (%)	Count	Percentage (%)	Count	Percentage (%)	Count	Percentage (%)	Count	Percentage (%)
D-D	0	0.0	0	0.0	0	0.0	0	0.0	0.4	1.1
M-D	0	0.0	0	0.0	0	0.0	0	0.0	0.0	0.0
R-D	46	90.2	23	67.6	35	74.5	21	77.8	32.2	76.7
C-D	5	9.8	11	32.4	12	25.5	6	22.2	9.3	22.2
Total	50	100.0	34	100.0	47	100.0	27	100.0	42	100.0

Table 9.6 The counts and percentages of each demonstration code in the *Relativism-Commitment* group

Codes	Ken		Kevin		Kirk		Average	
	Count	Percentage (%)	Count	Percentage (%)	Count	Percentage (%)	Count	Percentage (%)
D-D	0	0.0	0	0.0	0	0.0	0.0	0.0
M-D	0	0.0	0	0.0	0	0.0	0.0	0.0
R-D	14	43.8	41	66.1	29	78.4	28.0	62.8
C-D	18	56.3	21	33.9	8	21.6	15.7	37.2
Total	32	100.0	62	100.0	37	100.0	43.7	100.0

As shown in Table 9.5, the transcripts of most of the participants from the Relativism group demonstrated very few instances of dualistic and multiplistic thinking. Across all participants, they demonstrate around 60 % or more instances of relativistic thinking with most of the rest of the instances in commitment to relativistic thinking. On average, the percentage of dualistic and multiplistic thinking is close to zero. The average percentage of the codes in relativistic thinking is 76.7 %, and the average percentage of the codes in commitment to relativistic thinking is 22.0 %.

As shown in Table 9.6, the transcripts of the three participants from the *Relativism-Commitment* group showed no demonstrations of dualistic or multiplistic thinking. The percentage of the instances of relativistic thinking varies from 40 to 80 %. The percentage of the instances of commitment to relativistic thinking varies from 20 to 60 %. Similar ranges were observed for the participants in the group of *Commitment* (Table 9.7). On average, the percentage of dualistic thinking and multiplistic thinking for both groups is zero. The average percentage of the Relativistic Thinking-Demonstration codes is 63.6 % for the *Relativism-Commitment* group and 63.3 % for the *Commitment* group. The average percentage of the Commitment-Demonstration codes is 36.4 % for the *Relativism-Commitment* group and 36.7 % for the *Commitment* group.

The average percentages of the demonstration codes for the five groups (*Dualism, Multiplicity, Relativism, Relativism-Commitment,* and *Commitment*) are compiled as illustrated in Fig. 9.1. This quantitative representation of the qualitative data validates the groupings established from the survey results in the following ways:

First, on average, the *Dualism* group has a higher percentage of the Dualistic Thinking-Demonstration codes and lower percentage of the Relativistic Thinking-Demonstration and Commitment-Demonstration codes than all the other groups.

Second, the groups of *Relativism, Relativism-Commitment,* and *Commitment* all have few demonstration codes in dualistic thinking or multiplistic thinking. Their demonstration codes focus on relativistic thinking and commitment to relativistic thinking.

Table 9.7 The counts and percentages of each demonstration code in the *Commitment* group

Codes	Cameron		Charles		Charlie		Cody		Average	
	Count	Percentage (%)	Count	Percentage (%)	Count	Percentage (%)	Count	Percentage (%)	Count	Percentage (%)
D-D	0	0.0	0	0.0	0	0.0	0	0.0	0.0	0.0
M-D	0	0.0	0	0.0	0	0.0	0	0.0	0.0	0.0
R-D	49	84.5	23	65.7	40	78.4	17	31.5	32.3	65.0
C-D	9	15.5	12	34.3	11	21.6	37	68.5	17.3	35.0
Total	58	100.0	35	100.0	51	100.0	54	100.0	49.5	100.0

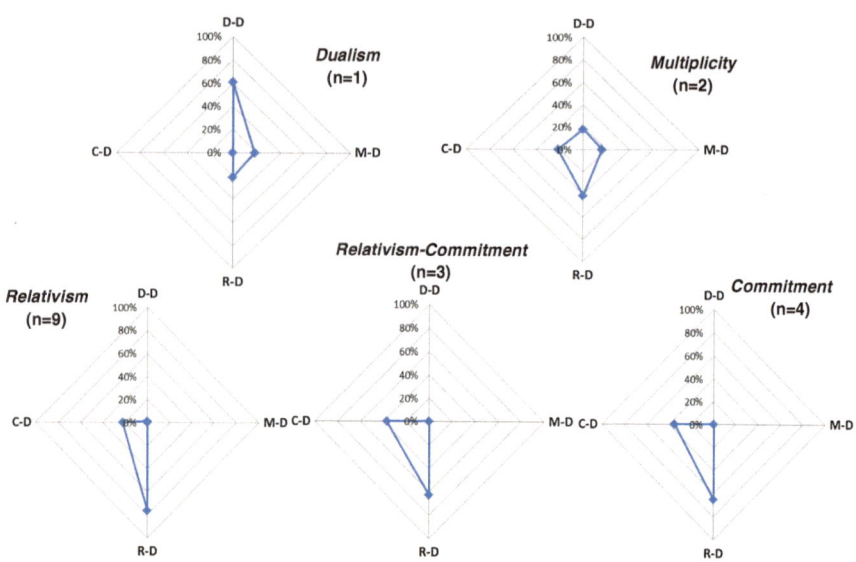

Fig. 9.1 A radar chart of the average percentages of the demonstration codes for the five different groups

Third, the only difference for the distributions of the demonstration codes across the groups of *Relativism, Relativism-Commitment,* and *Commitment* is that the *Relativism-Commitment* and *Commitment* groups seem to have higher representations of the Commitment-Demonstration codes than the *Relativism* group. This observation is consistent with the analysis based on the survey results. The naming of the group, i.e., *Relativism-Commitment* or *Commitment,* suggests the prominent thinking styles of that group. That is to say, the prominent thinking style of the *Commitment* group is the style of commitment to relativistic thinking. Therefore, it is reasonable that the percentages of Commitment-Demonstration codes for these two groups are higher than those of the *Relativism* group because they both have *Commitment* in their prominent thinking style(s).

Meanwhile, it should also be noted that, although for the *Commitment* group the prominent thinking style is supposed to be commitment to relativistic thinking according to the survey results, the higher percentage of the Relativistic Thinking-Demonstration codes than Commitment-Demonstration codes cannot be viewed as a conflicting result. This is because the stage of *Commitment* in Perry's theory essentially means a commitment to relativistic thinking. Therefore, it is still reasonable to see a higher representation of Relativistic Thinking-Demonstration codes than Commitment-Demonstration codes.

Overall, the observations drawn from the frequency counts and the distributions of qualitative coding results validated the groupings established from the quantitative round. Using the participants from all of the five groups as examples, the following sections will focus on the operationalization of each thinking style and the possible factors related to relativistic thinking. In the following sections, the cases of relativistic thinking and commitment to relativistic thinking constitute the major part of the qualitative research, in part because the majority of the students fell in the *Relativism, Relativism-Commitment,* and *Commitment* groups. The cases of dualistic thinking and multiplistic thinking are also described. However, because of the small representation of students in the *Dualism* and *Multiplicity* groups, the findings may not be as generalizable as the other two styles of thinking. Nonetheless, the information gathered from these two groups permits the picture of epistemological developmental profiles to be more complete. Despite the differences in significance of these findings to this current research, the findings are still presented in an order according to their original developmental order according to Perry's theory. This has been done to facilitate the understanding of this information. It should be noted that the emphasis of this research is on the operationalization of relativistic thinking and commitment to relativistic thinking and the factors associated with relativistic thinking.

9.3.2   The Story of Dualistic Thinking

Although only one participant, David from the group of *Dualism,* agreed to be interviewed, the qualitative interview with this participant still provides some valuable information about the operationalization of dualistic thinking among Chinese engineering doctoral students. Since nearly 60 % of the Dualistic Thinking-Demonstrations codes come from David's transcript with the rest from several other transcripts, David is used here as an example to illustrate the operationalization of dualistic thinking. A complete list of all of the Dualistic Thinking-Demonstration codes can be found in Appendix E.

A detailed breakdown of all the counts of the Dualistic Thinking-Demonstration codes for David is shown in Table 9.8. As is shown in this table, "Professors as the mediator of the right answers to questions or procedures" is listed as the top instance, and next to this code there are three codes with three counts each. These codes are "Authoritative figure is very knowledgeable or capable, higher than the students," "Diversity in opinion has been given a place," and "To learn the procedural knowledge/skills in doing research."

Table 9.8 Dualistic thinking-demonstration code list for David

Third-level code list	David
Professors as the mediator of the right answers to questions or procedures	9
Authoritative figure is very knowledgeable or capable, higher than the students	3
Diversity in opinion has been given a place	3
To learn the procedural knowledge/skills in doing research	3
To take order in a passive manner	1
Uncertainty has been given a place	1

As is shown in Table 9.8, "Professors as the mediator of the right answers to questions or procedures" is placed at the top in the list. As it states, this code refers to a student's regard of the professors (i.e., either his/her advisor, course instructors or other professors) as the mediators of the right procedures or the right answers to his/her question. Here is a quote to illustrate the code, "Professors as the mediator of the right answers to questions or procedures":

> *Um, I think they (feedback or suggestions from the advisor) are all very helpful to me. He, first of all, in the knowledge level, say, he will, when I encounter a problem, say, like, several days ago, I ran a simulation. He, oh, I was not very familiar with it (simulation). So, he gave me, like a simple, lecture, a few minutes, he talked to me about the codes, about how to write the codes, then, the rough programming, I mean, we needed to write programs, he gave me the pseudocode. After that, I knew how to write the codes, I had a rough idea about it, I wouldn't be like, completely lost. Yeah, first it is about learning the knowledge, he gives, mini-lectures once in a while.*

-David

Codes:

Dualistic Thinking - Demonstration - Professors as the mediator of the right answers to questions or procedures;

Dualistic Thinking - Demonstration - To learn the procedural knowledge/skills in doing research

This quote is also double-coded with "To learn the procedural knowledge/skills in doing research." In addition to regarding the professors as the mediator of the right answers to questions or procedures and learning the procedural knowledge/skills in doing research, the student also regards the authoritative figure (e.g., advisors, courses instructors, etc.) to be very knowledgeable or capable, at a level higher than that of the students. A quote to illustrate the code "Authoritative figure is very knowledgeable or capable, higher than the students" is shown as follows.

> *For my advisor, I think, my advisor is a very, very capable person…yeah, if you listen to (the presentations), two students (from our group) graduated this semester, one Ph.D. one Master, and then, not only him, but also the whole committee, you will find that, because, for the work done by the two students, I basically have no idea. But you can see that for the professors, although their research fields were not the same as the students' fields, they can always get the point. Moreover, some of their questions even challenged the students so much that the students cannot answer them. You found that, it seemed, they knew so much. Then, oh, (laughter) we, several students, we sat beside them, we were, like, at lost, we had just been there for nothing.*

> *I have asked other students, say, "Can you, you understand the work of the person?" He replied, "I basically have no idea." I think, including my roommate, their department, when they hold their group meeting, when the senior lab mates were giving presentations, he cannot understand either. But then you find that, they are amazing, the professors, they seem to understand everything.*

-David

Code:

Dualistic Thinking - Demonstration - Authoritative figure is very knowledgeable or capable, higher than the students

It seemed that the student had given "diversity in opinion" a place in his thinking, although he has personally suggested that he had never encountered such cases.

Yeah, at least so far I have never encountered two kinds of cases, two groups of people, one group says this one is good, and the other says that one is good, but, they cannot decide, and they don't know what method to use to determine. I probably haven't reached that high level, I guess.

-David

Code:

Dualistic Thinking - Demonstration - Diversity in opinion has been given a place

Since not many students were grouped into the group of *Dualism* according to the survey results ($n = 7$), the findings here provide only a snapshot of Chinese engineering doctoral students whose prominent thinking style is dualistic thinking. While a snapshot can assist in the assembly of a more complete picture of the students' epistemological developmental profile, the findings in this section may not be generalizable because of the small representation of the *Dualism* group among the survey respondents and the small number of interview participants ($n = 1$).

9.3.3   The Story of Multiplistic Thinking

Two participants from the *Multiplicity* group agreed to be interviewed. According to their responses to the survey, Mary showed a high value only in the *Multiplicity* subscale but not in the other subscales. The other student, Mike, showed high values in both the *Multiplicity* and *Relativism* subscales with the value in the *Multiplicity* subscale being of a higher significance level.

The qualitative interviews with these participants provide interesting and valuable information about the characteristics and demonstrations of multiplistic thinking among Chinese engineering doctoral students. It is especially interesting that in Mary's response to the first interview question, "What stands out in the last semester," she suggested that something happened a week before the interview. The time point of the event was right between when she finished the survey and the time when she was interviewed. Here is a quote that captured the attention of the interviewer right at the point of interviewing,

It was last week, when I met with him (the advisor), when I talked to him, because, before that, I did not quite know what is research, and before that, I had no idea what research was, I had never done real research before. But, when I was talking to him that day, I was, **maybe, it had been accumulated to a certain degree, and then, on that day, I suddenly realized,** *I realized in this field, what the progresses of other people are, and what kinds of* *things I can still do, basically, it allowed me to have some concepts of my field, of how to do research.*

-Mary

Code:

Relativistic Thinking - Demonstration - To be aware of what is going on in the field (with an explicit intent to inform one's own research or study)

Her expressions of her thinking processes have demonstrated some relativistic thinking already. This particular quote was coded with "Relativistic Thinking-Demonstration—To be aware of what is going on in the field (with an explicit intent to inform one's own research or study)." Her description of her experiences seemed to suggest that she was right at an "Ah-ha" moment about doing research. Her descriptions suggest a possible transition to the way of relativistic thinking.

With this interesting observation documented, this participant could have experienced some transitions between the time she took the survey and the time she was interviewed. As observed from the transcripts, none of the four thinking styles seemed to be Mary's prominent thinking in her transcript.

For the other participant, Mike, who scored high in both *Multiplicity* and *Relativism* in the survey, his transcript seemed to illustrate instances in relativistic thinking and commitment rather than in multiplistic and dualistic thinking. Nonetheless, he still showed a few instances of dualistic thinking and multiplistic thinking, which are often absent for the students in the *Relativism* and/or *Commitment* groups. This observation suggests that Mike could have gone further in his transitions in Perry's scale because of the multiple representations of different thinking styles.

Since most of the instances in multiplistic thinking still came from Mary's transcript, she is used here as an example to illustrate the operationalization of multiplistic thinking. A complete list of all Multiplistic Thinking-Demonstration codes can be found in Appendix F.

A detailed breakdown of all of the counts of the Multiplistic Thinking-Demonstration codes for Mary's transcript is shown in Table 9.9. As this table indicates, "To be aware of other students/faculty/researchers' opinions, ideas, progresses or other info without discernment" is at the top of the list, followed by "To collect information in a large amount or in an active manner WITHOUT explicit discernment" with three counts. Next in the list is "To discover the limitation of authority" with two counts. There are three codes with one count. These codes are "To accept the legitimacy of diversity in opinions," "To accept the legitimacy of the uncertainty of knowledge," and "To understand the evaluation of process in terms of the length of a paper/dissertation."

As shown in Table 9.9, "To be aware of other students/faculty/researchers' opinions, ideas, progresses or other info without discernment" listed at the top of the list. In essence, this code refers to a student who is aware of other

Table 9.9 Multiplistic thinking-demonstration code list for Mary

Third-level code list	Mary
To be aware of other students/faculty/researchers' opinions, ideas, progresses, or other information WITHOUT discernment	3
To discover the limitation of authority	2
To accept the legitimacy of diversity in opinions	1
To accept the legitimacy of the uncertainty of knowledge	1
To understand the evaluation of processes in terms of the length of a paper/dissertation	1

students/faculty/researchers' opinions, ideas, progresses, or other information, but without explicit discernment. A quote in this case is,

R: "... In this case, how do you make a decision, if there are different interpretations?"

E: "As a matter of fact, I think, there is no need to make some decisions, like, I decide to adopt this opinion or decide to adopt the other opinion. Actually I think, just to know about, the existence of these opinions, Um, before getting any other result, first, just to know about these other opinions, I think, it is enough."

-Mary

Code:

Multiplistic Thinking - Demonstration - To be aware of other students/faculty/researchers' opinions, ideas, progresses, or other information without discernment

(Note: R, interviewer; E, interviewee. These abbreviations are used for the quotes that involve both the interviewer and interviewee)

A quote to illustrate the idea of "To discover the limitation of authority" reads,

Sometimes when I was asking questions, I found that he was not very clear either. We then think maybe these areas were worth more discussions, worth researching. Um, but sometime, he, anyways, he still knows much more than I do, so, many times, he can provide me with a lot of information, or some directions.

-Mary

Code:

Multiplistic Thinking - Demonstration - To discover the limitation of authority

As shown in the quote, the student respects the advisor's scope of knowledge; however, she has started to notice that her advisor cannot always provide very clear answer. Here is another interesting code that illustrates that the student has also begun to accept the legitimacy of the diversity of opinions and the uncertainty of knowledge, elements that are the core ideas of *Multiplicity*:

Because now I am a graduate student, in the class, the things that the instructor talks about will be more uncertain, that is, because, the more defined ones, the results found by earlier researchers, we have already covered in the undergraduate studies, those are the things well established for now. What we are learning right now, are more, are the results done by more recent researchers, therefore, many times, it is not a single idea, say when you explain a certain thing, a certain phenomenon, there is not just one, one explanation, there

are, say, several people have one explanation, the others have some other explanation,
those times...I mean, in my current classes, there will be, what we are learning can have
the cases in which there are more than one answer.

-Mary

Codes:

Multiplistic Thinking - Demonstration - To accept the legitimacy of diversity in opinions

Multiplistic Thinking - Demonstration - To accept the legitimacy of the uncertainty of
knowledge

To summarize the stage of *Multiplicity*, students in this stage no longer regards authority figures as the mediators of all the right answers. This fact is evidenced by a decrease in the counts of demonstrations of dualistic thinking. These students have started to find some areas in which professors themselves cannot find clear answers or solutions to problems. Meanwhile, these students have started to show some demonstrations of the relativistic thinking. Due to the complexity of episte-mological development, the development should not be regarded as a linear process. Some students, especially those in a possible transitional stage, can often exhibit more than one type of thinking. Students who are classified as *Multiplicity* have exhibited quite a lot of instances of relativistic thinking and even commitment to relativistic thinking.

Similar to the *Dualism* group, not many students fell into the *Multiplicity* group according to the survey results ($n = 9$). Therefore, these findings provide only a snapshot of Chinese engineering doctoral students whose prominent thinking style is multiplistic. This snapshot helps to make the picture of epistemological developmental profile more complete, but the findings in this section may not be generalizable because of the small representation of the *Multiplicity* group and the small number of interview participants ($n = 2$).

9.3.4   The Story of Relativistic Thinking

In Perry's theory, a commitment refers essentially to a commitment to relativistic thinking; therefore, it is reasonable to also include the *Commitment* group in this discussion of the story of relativistic thinking. Nine students were classified in the group of *Relativism*, three were classified in the *Relativism-Commitment* group, and four in the *Commitment* group. The story of *Relativism* includes participants from all three groups ($n = 16$) since they all exhibit examples of relativistic thinking.

The total number of the Relativistic thinking-Demonstration third level codes within these three groups is 49. The Relativistic Thinking-Demonstration codes with counts equal to or larger than 5 across all three groups are illustrated in Table 9.10. A complete list of all Relativistic Thinking-Demonstration codes can be found in Appendix G.

Table 9.10 Relativistic thinking—demonstration codes list for the groups of *Relativism*, *Relativism-Commitment* and *Commitment*

Third level codes	Rick	Robert	Ron	Rena	Ryan	Ruby	Rose	Rebekah	Ray	Ken	Kirk	Kevin	Cameron	Cody	Charles	Charlie	Total	Count of students
1. To regard professors/advisors' role as a guide	3	3	1	4	2	1	3	4	4	2	4	0	4	1	6	5	47	15
2. To actively ask others questions or seek out help	0	4	1	2	1	2	0	3	1	1	0	5	3	1	5	7	36	13
3. To collect information in a large amount or in an active manner WITH explicit intent of discernment	4	3	3	0	5	0	1	1	1	0	2	4	3	1	1	3	32	13
4. To appreciate feedback from others	0	2	0	2	3	3	0	7	0	3	0	2	5	1	0	3	31	10
5. To find and/or solve problems	1	3	3	1	3	0	0	2	1	0	2	2	4	0	2	4	28	12
6. To discern information (e.g., papers, lectures, reports, etc.)	3	3	2	2	3	1	1	0	0	1	0	1	4	0	1	3	25	12
7. To take and/or test the feedback from others	2	4	4	2	4	0	0	2	3	1	0	0	1	1	0	1	25	11
8. To learn from others (generic)	1	3	6	0	1	0	3	0	0	3	0	3	0	3	0	0	23	8
9. To conduct or to anticipate conducting research in an independent manner	1	0	4	1	4	0	0	0	0	1	1	1	2	4	2	1	22	11
10. To learn in an independent manner	2	2	2	1	1	0	1	0	0	1	3	1	0	1	3	3	21	12
11. To defend one's positions/ideas and/or convince others	2	0	1	3	1	4	1	1	1	0	0	3	2	1	0	0	20	11
12. To think in an independent manner	1	4	1	4	3	0	0	0	0	0	2	0	4	0	0	0	19	7
13. To collaborate with others/to have teamwork skills	0	1	4	0	0	1	1	2	2	0	2	0	3	0	0	0	16	8

(continued)

Table 9.10 (continued)

Third level codes	Rick	Robert	Ron	Rena	Ryan	Ruby	Rose	Rebekah	Ray	Ken	Kirk	Kevin	Cameron	Cody	Charles	Charlie	Total	Count of students
14. To explore and test the unknown	1	1	2	1	2	0	2	0	0	0	0	2	2	0	0	1	14	9
15. To master knowledge at a deep level	1	2	0	0	1	0	0	0	0	0	2	3	1	0	1	3	14	8
16. To actively express one's idea (oral)	1	4	1	0	0	3	1	0	0	0	3	0	0	0	0	0	13	6
17. To appreciate conflicts and/or the diversity of opinions	1	2	1	0	0	0	2	0	1	0	1	1	0	1	1	1	12	10
18. To be aware of what is going on in the field (with an explicit intent to inform one's own research or study)	0	2	2	0	0	0	0	0	0	0	0	1	1	2	0	1	9	6
19. To expand one's scope of knowledge (with the intention to inform one's research/study)	0	1	0	0	0	0	2	0	1	0	2	1	2	0	0	0	9	6
20. To communicate with people from diverse backgrounds	0	2	1	0	0	1	1	0	2	0	0	0	0	0	0	0	7	5
21. To appreciate the deep thinking/philosophy of professors/advisor	0	0	0	0	0	0	0	4	0	0	1	0	0	0	0	1	6	3
22. To be aware of other students/faculty/researchers/experts' opinions, ideas, progress or other info with discernment	0	2	0	0	0	0	0	0	1	0	0	2	0	0	0	1	6	4
23. To produce sound research results	0	0	1	1	0	0	0	0	2	0	1	0	1	0	0	0	6	5
24. To learn to compromise	0	0	0	0	0	1	1	0	1	1	0	0	1	0	0	1	6	6

The Relativistic Thinking-Demonstration third-level codes shown in this table represent the operationalization of relativistic thinking among Chinese engineering doctoral students. In order to permit a better sense of the operationalization of the relativistic thinking, these codes are grouped into two categories: Communications/Interactions with Others and One's Own Skills in Research or Learning, along with their sub-categories (Table 9.11).

Since most of the instances of relativistic thinking came from the transcripts of all three groups (i.e., *Relativism*, *Relativism-Commitment*, and *Commitment*), in the following discussions, quotes from all of these groups are used as examples to illustrate the operationalization of relativistic thinking. Quotes that are typical examples of specific codes were chosen to illustrate the meaning of that particular code.

Interactions with Others

Interactions with professors/advisors
Compared to the ideas of students from the *Dualism* group, in which the students regard their professors/advisors as the mediators of correct answers or solutions, and compared to the ideas of students from the *Multiplicity* group, in which the students begin to recognize the limitations of their professors/advisors, here, most student (15 out of 16) have discussed their professors/advisors' role as one of a guide in their

Table 9.11 Two main categories for relativistic thinking-demonstration codes

Interactions with others	One's own skills in research or learning
Interactions with professors/advisors • To regard professors/advisors' role as a guide • To appreciate the deep thinking/philosophy of professors/advisor *Arguing with others* • To defend one's positions/ideas and/or convince others • To actively express one's idea (oral) • To learn to compromise *Learning from others* • To actively ask others questions or seek out help • To appreciate feedback from others • To take and/or test the feedback from others • To learn from others (generic) • To collaborate with others/to have teamwork skills • To appreciate conflicts and/or the diversity of opinions • To communicate with people from diverse backgrounds	*Learning/thinking/conducting research independently* • To learn in an independent manner • To think in an independent manner • To conduct or to anticipate conducting research in an independent manner • To collect information in a large amount or in an active manner WITH explicit intent of discernment • To discern information (e.g., papers, lectures, reports, etc.) • To be aware of what is going on in the field (with an explicit intent to inform one's own research or study) • To expand one's scope of knowledge (with the intention to inform one's research/study) • To be aware of other students/faculty/researchers/experts' opinions, ideas, progress, or other information with discernment *Other research/learning skills* • To find and/or solve problems • To explore and test the unknown • To master knowledge at a deep level • To produce sound research results

research or learning while the students themselves are often the ones who actually conduct the research in a more independent manner. Students are no longer looking up to their professors/advisors for the correct answers. They not only recognize the limitations of professors/advisors, but they also regard professor/advisors as collaborators, in part by respecting the ideas of these professors/ advisors and yet reserving their own right to discern the accuracy of researched information.

Here is an example to illustrate how Rebekah learned from her advisor "the philosophy of doing research":

> *It's from the change in my philosophy. That is, like, I talked to you about, about the change of my ideas from thinking of doctoral study as a "task" to now, as a journey of a "searcher;" in addition, it's an on-going searching, and that is why it's called a "re-searcher." It is such a process. Therefore, yeah, that is, in fact, I was talking with him (the advisor), you heard him saying these things, but you did not understand. By and by, little by little, you got it, and then you reached this conclusion. So, I think he is an inspirer.*

-Rebekah

Codes:

Relativistic Thinking - Demonstration - To regard professors/advisors' role as a guide

Relativistic Thinking - Demonstration - To appreciate the deep thinking/philosophy of professors/advisor

Relativistic Thinking - Factor - Advisor

Although they do regard their advisors as a guide, students with relativistic thinking often demonstrate an awareness to test the feedback they gather from these advisors. Here is an example from Rick about his interactions with his advisor:

> *E: Um, what I mean is, say, to do this project, usually I will try to solve the problems using my own ways. I will choose the direction that I want to go. But then, it's like, I will go and discuss with the advisor once in a while, let him decide, say, what I am doing, the direction that I want to go, whether that is a good direction or not for him. Of course he may have some different opinions with me. Then, we will have some debate and discussions. I mean, I will not always regard his opinions as 100 % correct. I will read a lot of materials, and discern for myself to see whether his opinions or my opinions are correct.*
>
> *R: Ok, thanks.*
>
> *E: But still it is a process to learn from the advisor, he is still the expert of the field. Also, he goes to conferences, so he has a broad view.*

-Rick

Codes:

Relativistic Thinking - Demonstration - To collect information in a large amount or in an active manner WITH explicit intent of discernment

Relativistic Thinking - Demonstration - To conduct or to anticipate conducting research in an independent manner

Relativistic Thinking - Demonstration - To discern information (e.g., papers, lectures, reports, etc.)

Relativistic Thinking - Demonstration - To find and/or solve problems

Relativistic Thinking - Demonstration - To take and/or test the feedback from others

Relativistic Thinking - Demonstration - To think in an independent manner

Commitment - Demonstration - To be the lead of oneself (life, goal, research, learning)

On one hand Rick has acknowledged the expertise of his advisor; on the other hand, he reserved the right to discern the feedback from his advisor. Moreover, Rick seemed to begin to take the lead in his own research, which in this research is regarded as a demonstration of a commitment to relativistic thinking.

In a manner similar to Rick's, Ron has described the role of his advisor as a guide in his research while he, himself, was the one to find the problems with the directions provided by his advisor.

> *I don't know about other students' professors; for our professor (advisor), he doesn't care about the details, but he will tell you about the status of the development in this field, he will do this kind of research and tell you what the overall direction is. After a while, (he will tell you) in which directions you need to submit the next paper, and what the current competitions are. It is YOU who should tell him, your opinions, after seeing all of the research problems, now, what your main problems are. Um, then he knows, "Oh, that is the case, but the other university has done these other things." So it's a process of on-going evaluating, on-going discussions.*

-Ron

Codes:

Relativistic Thinking - Demonstration - To regard professors/advisors' role as a guide

Relativistic Thinking - Demonstration - To find and/or solve problems

Relativistic Thinking - Factor - Advisor

Arguing with others

In addition to the changes in their interactions with the advisors, students with relativistic thinking seem to actively express their ideas. Many of them (11 out of 16) spoke about how they defended their positions/ideas or tried to convince others. Some mentioned learning how to compromise in their interactions with others. Some students stressed the importance of actively expressing their individual ideas.

As for the process of convincing others, it is interesting to see some of the convincing processes that have happened between these students and advisors. Here is an example that occurred with Ron:

> *So, hm, but, later, I still thought of my own idea. I did not tell him (the advisor) right after I had the idea. First I produced the results, produced the results according to my ideas and then showed the results to him. He saw the results. Indeed, the things were more valuable than what he wanted me to do. So I think, the process of convincing, I feel, it's, right, I feel very, (He did not finished the sentence. He nodded).*

-Ron

Codes:

Relativistic Thinking - Demonstration - To conduct or to anticipate conducting research in an independent manner

Relativistic Thinking - Demonstration - To defend one's positions/ideas and/or convince others

These convincing processes have happened not only with the students' advisors, but often also with their classmates, especially when these students were involved in some group projects and had the need to discuss these projects with their classmates. For example, Rick said:

Usually, for classmates, with some homework, or some, because the professors here will assign presentations, these presentations are done in groups, so there will be some big groups, there will be opportunities to communicate with them. Often we discuss some problems, I mean, because everyone has their own ideas, so I may need to find ways to convince others. Yeah, these are my communications with my classmates.

-Rick

Codes:

Relativistic Thinking - Demonstration - To defend one's positions/ideas and/or convince others

Relativistic Thinking - Factor - Discussions with other students

Relativistic Thinking - Factor - Experiences with group projects

The experience of convincing others has often taken place hand-in-hand with the experience of learning to compromise. Here is what Rose described about her discussions in class:

E: In the class, we will discuss with each other. For example, in the beginning, we need to decide a project topic through our discussions. Sometimes we need to compromise too. For example, (if I say) "Oh, I think that is a good topic." But all of the other students like some other topics, then you need to compromise, or say, because we were in a discussion, there was feedback, that was how you know whether others were interested in your topic or not. If you know that they are not interested in the topic that you are interested in, you need to try to convince others. I mean, if you are alone, no feedback, there is no need for convincing others, as long as you yourself are interested, then you are fine. If there are many people, then you need to convince others, you need to state the reasons, no. 1, 2, and 3, why this topic is better than the other ones, or, say, I know quite a lot about this topic, so these are the advantages if we choose this topic. I think, when I was in the class, first, I need to compromise; Second, I need to convince others.

-Rose

Codes:

Relativistic Thinking - Demonstration - To defend on's positions/ideas and/or convince others

Relativistic Thinking - Demonstration - To learn to compromise

Relativistic Thinking - Factor - Experiences with group projects

Relativistic Thinking - Factor - Discussions with other students

In addition to the experiences of convincing others or being convinced, students often spoke about the importance of actively expressing one's ideas to professors or classmates:

> *I mean, you should talk more to the professors. You should let him know your thoughts regularly. In this case, they will know what you are doing. Or, say, they can give you advice. But if, if you focus on your own stuff, maybe, say, you speak up when you encounter a problem, then it will probably be too late. Therefore, you should always update your thoughts, or your future plans. It is very important.*

-Robert

Codes:

Relativistic Thinking - Demonstration - To actively express one's idea (oral)

In sum, the presence of these codes suggests the demonstration of relativistic thinking in the form of students' active involvement in an argument. The processes of actively expressing one's own ideas, convincing others, or being convinced indirectly imply the students' relativistic thinking in terms of their ability to weigh and evaluate different evidence, opinions, and solutions in these processes.

Learning from others

Another main demonstration of students with relativistic thinking is found in their actively learning from others. Most of them (13 out of 16) discussed the importance of actively asking questions or seeking out help from others. Also, 11 out of 16 mentioned that they take and/or test the feedback from others. Many of them (10 out of 16) have started to appreciate the value of feedback from others. Some other formats of learning from others include to collaborate with others/to have teamwork skills, to appreciate conflicts and/or the diversity of opinions, and to communicate with people from diverse backgrounds. Eight out of 16 of these studied individuals spoke about learning from others in a generic sense. The prominent presence of this category of demonstration code suggests that students with relativistic thinking have come to understand the significance of diverse views and opinions, actively exposed themselves to opportunities to obtain helpful feedback, and practiced the process of testing and evaluating opinions and feedback from others.

Here is a quote from Rebekah in which she spoke about the importance of actively asking others questions or seeking out help when she was writing a draft for her qualifying examination.

> *E: Yeah. Many people from the research lab helped, to revise, from the outline, that is, after I drafted the outline, I gave it to others for comments. After I finished a draft, I sent it out, to many people for help. I really appreciate their help; they gave me a lot, lots of honest feedback. Yeah, one draft after another, I felt, from the first draft to the last draft, it was completely different. Right.*
>
> *R: You said honest feedback, what types of feedback do you mean?*

E: All kinds of, as big as the structure. Structurally, they gave a lot of suggestions. Especially in the first draft, they provided many comments. (In the draft, there were) Some, confusions. So, that is, really, it was changed a lot structurally. Small things like, grammar. Yeah, step by step. So I felt, really, I appreciated their help.

-Rebekah

Codes:

Relativistic Thinking - Demonstration - To actively ask others questions or seek out help

Relativistic Thinking - Demonstration - To appreciate feedback from others

Relativistic Thinking - Demonstration - To take and/or test the feedback from others

Relativistic Thinking - Factor - Senior lab mates or school mates

From this quote, it appears that in addition to Rebekah's attempt to actively seek out help, she has also developed an appreciation for the feedback obtained from others. These students not only appreciate the feedback, but they also described the testing of feedback. The testing of feedback represents the core ideas of relativistic thinking in the evaluation and weighing of evidences. Ryan expressed his thinking process in the following terms:

For the disadvantages, first, you will definitely need to test, whether the disadvantage he (the advisor) points out is really a disadvantage or not. Right, this, because what he says may not be correct. It's just his thought. Therefore, you need to test this thing. Um, but, when you are going through the process, what he mentioned, say, places that need special attention, you need to also pay attention to them. You should check, say, whether his opinion is correct or not, whether the factor is important or not.

-Ryan

Codes:

Relativistic Thinking - Demonstration - To take and/or test the feedback from others

In addition to taking and testing feedback from others, some students with relativistic thinking have demonstrated an appreciation of conflicts of opinions or the diversity of opinions. Here is an example from Charles in which he has recognized the value of having conflicts of ideas:

Um, there will be different perspectives of thinking. Putting together the thoughts from everyone, there could be some sparks. Um, thinking only by oneself, the scope can be narrow. But, everyone, if putting everyone's ideas together, even though some ideas may not be very reasonable, still it can broaden your thinking.

-Charles

Codes:

Relativistic Thinking - Demonstration - To appreciate conflicts and/or the diversity of opinions

Relativistic Thinking - Factor - Discussions with other students

Here is an example from Ray in which he has recognized the value of having conflicts of ideas while conducting his research:

...If it is a problem about designing a system, we would, try it first. If we have conflicts among us, actually, I think it is a good thing, good, actually it is good. That is, if there are conflicts in our opinions, it's easy, 'cause, say the student or the teacher, or among students, one side can be convinced, if there are data. Another good thing is, if none of them gave up their ideas, then, you need to try, I would try the ideas of both sides, to see which one works. It can help research, I think, that how we do things.

-Ray

Codes:

Relativistic Thinking - Demonstration - To appreciate conflicts and/or the diversity of opinions

Relativistic Thinking - Demonstration - To take and/or test the feedback from others

In summary, relativistic thinking of students is reflected in their interactions with their professors/advisors, classmates, lab mates, and/or other researchers. In the students' interactions with others, they have often undertaken the processes of independently evaluating opinions and weighing evidence. These processes are reflected in their appreciations of their professors/advisors' philosophy, feedback from others, and conflicts of opinions, and in their own abilities for testing the feedback.

One's Own Skills in Research or Learning

Learning/Thinking/Conducting Research Independently
Most students spoke about the importance of learning in an independent manner (12 out of 16). Many students discussed their doing research or the anticipation of doing research in an independent manner (11 out of 16). Some emphasized the importance of thinking in an independent manner (7 out 16).

Here is an example taken from Rick, as he spoke about his ideas of independent learning:

I mean, um, I think learning is not through, I mean, it requires one to actively seek, to actively read a lot of materials, then you can master the knowledge. If you only learn by attending classes, through others passing the knowledge onto you, I mean, first of all, it is very difficult to learn in that way; also, you may forget the content of the classes pretty soon. So I think learning is an autonomous process.

-Rick

Codes:

Relativistic Thinking - Demonstration - To learn in an independent manner

Relativistic Thinking - Demonstration - To collect information in a large amount or in an active manner WITH explicit intent of discernment

In addition to referring to learning in an independent manner, students have discussed conducting research and thinking in an independent manner. Ryan expressed his ideas about the importance of independent thinking and doing research in an independent manner.

Being independent, autonomy is the most important. So, many people, many people grad-uated from very good labs, very very good labs, the advisors were excellent. After they graduated, they did research, they became independent faculty members, but the outcomes were really bad. That's because in the training process, he did what the professors told them to. He never has his own ideas, his independent ideas, so he did not develop this independent research ability. Right, this is very important. Because I do know some people, someone from a biology major. They did a lot of experiments. Say the advisor says you do these tasks this week, the other tasks next week. But after they graduated and became professors, they needed to write proposals, to obtain funds, and they had no idea what to write about.

-Ryan

Codes:

Relativistic Thinking - Demonstration - To think in an independent manner

Relativistic Thinking - Demonstration - To conduct or to anticipate conducting research in an independent manner

In sum, the theme of being independent in one's own thinking, learning, and research has been prevalent across the participants from the group of *Relativism, Relativism-Commitment*, and *Commitment*. It seems that, on one hand, students with relativistic thinking regard the professors/advisors' role as a guide to be important, while, on the other hand, these students are also actively striving to learn or think independently.

Discerning of Information

In addition to thinking/learning/conducting research in an independent manner, one of the most direct representations of relativistic thinking is the ability to discern information. This representation is evidenced in students' experiences in discerning different kinds of information, such as academic papers, lectures, presentations, and even online resources. Among 16 students, 12 of them talked about their experiences in discerning different kinds of information. As compared to the representations of multiplistic thinking in which students accept the legitimacy of diversity of opinions, here, the students with relativistic thinking often highlight an explicit process of discernment when facing diverse information and resources. Moreover, the skills or abilities described are featured to illustrate the students' focused intent for informing their own research or study. These skills or abilities include to collect a large amount of information or to collect information in an active manner, to be aware of what is going on in the field, to expand personal scopes of knowledge, and to become aware of the other students/faculty/researchers/experts' opinions, ideas, progress, or other information.

For the purpose of discerning information, students spoke of the discernment of information in different formats. Here is an example of Rena discussing her experiences with professional conferences.

Maybe, a while ago, I attended some academic conferences, when I listened to some other researchers presenting their work. I had a lot of questions. For some of the stuff, I was wondering, well, was it that easy? Did they complete the experiments in such a simplified manner? Can this result stand questioning? Yeah, we questioned (their results).

-Rena

Codes:

Relativistic Thinking - Demonstration - To discern information (e.g., papers, lectures, reports, etc.)

Relativistic Thinking - Factor - Professional conferences

Here is another example from Ron, who spoke of discerning in terms of academic papers:

When I was in undergraduate study, um, because, there is only one answer (to a question), so I often relied on the textbook, I relied upon past experiences. It's enough to just go and get to, just to figure out the correct answer. But for research, you need to read a lot of papers. As a matter of fact, some content of the papers are correct, some are incorrect, you need to discern yourself.

-Ron

Codes:

Relativistic Thinking - Demonstration - To collect information in a large amount or in an active manner WITH explicit intent of discernment

Relativistic Thinking - Demonstration - To discern information (e.g., papers, lectures, reports, etc.)

The discerning of information often occurs hand in hand with the collection of large amounts of information. Here is another example:

Sometimes the teacher would say, you should, you should do some literature search, on this topic. For me, I will first check if, first I will read some textbook to see if there is any classic theory in this area. Secondly, I will check, because classic theories are often dated, for example, and I want to know, now, now, in academia or in industry, what the perspectives towards this topic are. I will then search more recent (literature) reviews, then using the reviews to check their references, then I can roughly get the idea about the history and up till now, this topic, how projects in this topic evolved, like, in research, what their opinions are, that is, compared to the past research, what the latest progress is, what the main problems are. If I were to do this topic, what are the areas that I should expand upon, and go deeper.

-Rick

Codes:

Relativistic Thinking - Demonstration - To collect information in a large amount or in an active manner WITH explicit intent of discernment

Relativistic Thinking - Demonstration - To discern information (e.g., papers, lectures, reports, etc.)

Relativistic Thinking - Demonstration - To explore and test the unknown

Students have tried to stay informed about the recent progress of the field with a clear intent to inform one's own research. Here is how Robert spoke about his experiences.

...You should make more use of the resources, I mean, if the university is relatively large, especially if the engineering school is relatively large, then you will have more opportunities to get in contact with the information from your own field, or from your interdisciplinary fields, it will give you, it will stimulate your innovative thinking. That's right, you should be exposed to more of these types of opportunities, such as seminar lectures, I think, it's good to just go and listen.

-Robert

Codes:

Relativistic Thinking - Demonstration - To be aware of what is going on in the field (with an explicit intent to inform one's own research or study)

Some students have also discussed the process of expanding their own knowledge scope for the purpose of research itself. Here is how Rose talked about it:

You can think in a different way: all the knowledge is there, whether I learn about it or not makes no difference to this world, I mean, it makes no difference to the knowledge itself. So, the reason that causes you to learn is due to the fact that this particular knowledge is new, is unknown to me. If you have this type of attitude, then you can do research! It's like, everything is new to you, although it may not be new to others. But, if you keep searching, one day, you will find one thing that is new to you, and new to other. That is the way.

-Rose

Codes:

Relativistic Thinking - Demonstration - To expand one's scope of knowledge (with the intention to inform one's research/study)

Relativistic Thinking - Demonstration - To explore and test the unknown

Rose's process of researching new knowledge is accompanied with her intention to explore and search the unknown. The learning and discerning of knowledge and/or other types of information are closely related with other research and/or learning skills, which will be discussed in the following section.

Other Research/Learning Skills

Other important skills in doing research and learning include to find and/or solve problems (12 out 16), to explore and test the unknown (9 out of 16), to master knowledge at a deeper level (8 out of 16), and to produce sound research results (5 out of 16).

Here is an example from Ron as he discussed developing the ability to debug.

Um, also, I continue to develop the ability to debug. I mean, I need to know not only whether there is a problem or not, but also, where the problems are, and how to find the problem and how to solve the problem.

-Ron

Codes:

Relativistic Thinking - Demonstration - To find and/or solve problems

Some students described the process of exploring and testing the unknown. Here is what Charlie commented on regarding the process of doing research:

Of course, there are things you have to try before you know whether it is a good fit or not. Once you encounter some problem, you probably need to stop immediately and try something else. But, it is also possible that it is a good fit, but if you just haven't done it thoroughly, this is possible too.

-Charlie

Codes:

Relativistic Thinking - Demonstration - To explore and test the unknown

Other students commented on the importance of "really learning something," that is, to master certain knowledge at a deep level instead of merely learning something in a superficial manner.

Because it's different for graduate students and undergraduate students, graduate students focus on really learning something. So, basically, you are trying to figure out something, to figure out what this thing really is, how it is done. Basically this is the case, it's different from undergraduate students. In undergraduate studies, it's more for the grades.

-Cameron

Codes:

Relativistic Thinking - Demonstration - To master knowledge at a deep level

To summarize, the skills or abilities discussed to this point provide further examples of relativistic thinking within Chinese engineering doctoral students' research and learning. These demonstrations, along with the above-mentioned abilities, including independent learning/thinking/doing research and collecting and discerning information, provide different concrete representations of relativistic thinking among these students.

9.3.5 The Story of Commitment to Relativistic Thinking

Many students from the *Relativism* and *Relativism-Commitment* groups have demonstrated some level of commitment to relativistic thinking. Therefore, it is reasonable to also include the groups of *Relativism*, *Relativism-Commitment*, and *Commitment* in this discussion. Nine students were classified in the group of *Relativism*, three were classified in the *Relativism-Commitment* group, and four in the *Commitment* group. The story of committing to relativistic thinking includes participants from all three groups ($n = 16$) since they all exhibit a certain level of commitment to relativistic thinking.

The total number of the Commitment-Demonstration third level codes across these three groups is 55. The Commitment-Demonstration codes with counts no less than three across all three groups are shown in Table 9.12. A complete list of all Commitment-Demonstration codes can be found in Appendix H.

The Commitment-Demonstration third level codes shown in the table represent the operationalization of a commitment to relativistic thinking among Chinese

Table 9.12 Commitment-demonstration code list for the *Relativism*, *Relativism-Commitment*, and *Commitment* groups

Third level codes	Rick	Robert	Ron	Rena	Ryan	Ruby	Rose	Rebekah	Ray	Ken	Kirk	Kevin	Cameron	Cody	Charles	Charlie	Total	Count of students
1. To make a decision balancing different tensions	2	0	2	1	2	1	2	0	2	2	1	1	2	0	3	1	22	13
2. To take the major responsibilities in research or learning	0	1	0	0	0	0	0	0	0	1	0	2	1	8	0	1	14	6
3. To understand the uncertainty/limitations/implications of making decisions	1	5	0	0	1	0	0	0	0	1	1	0	1	0	0	3	13	7
4. To have a clear goal	0	0	0	0	1	1	0	0	2	3	0	1	0	1	1	2	12	8
5. To be the lead of oneself (life, goal, research, learning)	1	0	0	0	1	0	0	0	1	0	0	0	1	5	0	0	9	5
6. Anticipation of being the lead in one's career, research, work, etc.	0	1	0	0	0	2	0	0	0	1	0	2	1	0	0	0	7	5
7. Stylistic attributes (to do practical research vs. to do basic research)	0	0	1	1	0	0	0	1	2	1	0	1	0	0	0	0	7	6
8. To prepare for future career	0	0	0	0	0	0	0	0	0	3	2	1	0	0	1	0	7	4
9. To be persistent despite uncertainty	0	0	0	0	0	0	0	2	0	0	0	2	0	2	0	0	6	3
10. To not depend on the advisor or instructors	0	0	0	0	0	0	1	0	0	0	0	0	1	4	0	0	6	3
11. Stylistic attributes (learning/doing research out of real interest vs. learning/doing research for the sake of doing it)	1	1	0	0	1	1	1	0	0	0	0	0	0	0	0	0	5	5
12. To be brave/to overcome challenges	0	0	0	0	0	0	0	1	0	0	0	2	0	1	0	1	5	4
13. To be self-motivated	1	0	0	0	0	0	0	0	0	0	0	1	0	3	0	0	5	3
14. To do life-planning	0	0	0	0	0	0	0	0	0	1	0	0	0	2	1	1	5	4
15. To have a positive attitude	0	0	0	0	0	0	1	2	0	0	0	0	0	2	0	0	5	3
16. To have a strong motivation	0	1	0	0	1	0	0	1	0	0	0	0	0	0	0	2	5	4
17. Stylistic attributes (to be social professionally vs. to be isolated)	0	0	0	0	0	0	0	0	0	0	2	0	0	2	0	0	4	2
18. Stylistic attributes (to live a rich life vs. to live by essentials)	0	0	0	0	0	0	1	0	1	1	0	0	1	0	1	0	4	4

(continued)

Table 9.12 (continued)

Third level codes	Rick	Robert	Ron	Rena	Ryan	Ruby	Rose	Rebekah	Ray	Ken	Kirk	Kevin	Cameron	Cody	Charles	Charlie	Total	Count of students
19. To recognize the need to prioritize	0	1	0	0	0	0	0	0	0	0	0	3	0	0	0	0	4	2
20. Stylistic attributes (to challenge oneself vs. to choose the easy way out)	0	0	0	0	0	0	2	0	0	1	0	0	0	0	0	0	3	2
21. To be devoted to the work	0	0	0	0	0	0	0	0	0	0	0	1	0	1	1	0	3	3
22. To commit to a certain type of faith	0	0	1	1	0	0	0	0	0	0	0	0	0	1	0	0	3	3
23. To transform one's way of life because of his/her faith	0	0	1	1	0	0	0	1	0	0	0	0	0	0	0	0	3	3

Table 9.13 Five main categories for commitment-demonstration codes

Taking major responsibilities	Decision-making
• To take the major responsibilities in research or learning • To be the lead of oneself (life, goal, research, learning) • Anticipation of being the lead in one's career, research, work, etc.	• To make a decision balancing different tensions • To understand the uncertainty/limitations/implications of making decisions
Life planning	Stylistic issues
• To have a clear goal • To prepare for future career • To do life planning • To recognize the need to prioritize	• Stylistic attributes (to do practical research vs. to do basic research) • Stylistic attributes (learning/doing research out of real interest vs. learning/doing research for the sake of doing it) • Stylistic attributes (to be social professionally vs. to be isolated professionally) • Stylistic attributes (to live a rich life vs. to live by essentials) • Stylistic attributes (to challenge oneself vs. to choose the easy way out)
Attributes	Inner strength/faith
• To be persistent despite uncertainty • To be brave/to overcome challenges • To be self-motivated • To have a positive attitude • To have a strong motivation • To be devoted to the work	• To commit to a certain type of faith • To transform one's way of life because of his/her faith

engineering doctoral students. To have a better sense of the operationalization of the commitment to relativistic thinking, these codes are grouped into five categories (Table 9.13).

Since most of the instances of commitment to relativistic thinking came from the transcripts from all three groups (i.e., *Relativism*, *Relativism-Commitment*, and *Commitment*), in the following discussions, quotes from all of these groups are used as examples to illustrate the operationalization of commitment to relativistic thinking. Quotes that are typical examples of specific codes were chosen to illustrate the meaning of that particular code.

Taking Major Responsibilities

Students with relativistic thinking often regard their professors/advisors' role as that of a guide in their research or learning, and these students are the ones who actually conduct research in a more independent manner. Here, students start to take major responsibilities for their research, learning, and careers. These students talk about being the leader of oneself (life, goal, research, or learning) or their anticipation of taking the lead. Some of the students have described the actual process of doing research as not depending on their advisors. Instead, the students described the process of doing research as depending on themselves.

Here is an example showing Cody's experiences with doing research:

Ah, I see, my advisor, I would say, to be honest, because I am really doing pretty much all by myself, my advisor, uh, I am not sure, if my case is unique or not, but I don't think my advisor really plays that much role, I only meet her, update her, keep her updated about what is going on with me, but I am really doing all by myself, pretty much. 'Cause I guess that's the difference between a PhD and a Master. When I did my Master, actually, my advisor, she told me "Oh, what you should do." I just followed her steps, her instructions, but right now, I am pretty much, I design everything all by myself, I do everything myself, I don't think she does, (laugh), I am not downplaying her role, but, I mean, I am pretty much all depending on myself now.

-Cody

Codes:

Commitment - Demonstration - To be the lead of oneself (life, goal, research, learning)

Commitment - Demonstration - To not depend on the advisor or instructors

Commitment - Demonstration - To take the major responsibilities in research or learning

Relativistic Thinking - Demonstration - To conduct or to anticipate conducting research in an independent manner

Relativistic Thinking - Factor - Being in a doctoral program

And it is a similar case for Ryan:

Because when I was in China, I basically did all the research by myself, my case is more special. I did basically all of the stuff. For example, to find the research direction, to search the literature, to find the research problem, you can say I developed independent research skills.

-Ryan

Codes:

Commitment - Demonstration - To be the lead of oneself (life, goal, research, learning)

Relativistic Thinking - Demonstration - To collect information in a large amount or in an active manner WITH explicit intent of discernment

Relativistic Thinking - Demonstration - To conduct or to anticipate conducting research in an independent manner

Relativistic Thinking - Demonstration - To find and/or solve problems

The action of taking major responsibilities is often accompanied by the process of making a decision balancing different tensions, a process that will be discussed in the next section. Here is an example from Cameron:

For example, say, because the boss (advisor) is NOT the one who actually does the work, well, he may have some good, big ideas. But you will find those ideas are difficult to implement. So, so, then, you need to decide, you need to decide whether or not you will do it. If you do, to what degree you are going to do it. Well, but, it's like, if your boss happens to know quite a bit about the direction, you'd better follow him. For me, my boss happens to be just, interested in the direction, but he does not know a lot about it.

-Cameron

Codes:

Commitment - Demonstration - To make a decision balancing different tensions

Commitment - Demonstration - To take the major responsibilities in research or learning

Relativistic Thinking - Demonstration - To know the ability of the advisor in a given field

Decision-Making

In addition to taking major responsibilities, most of the students in all three groups (13 out of 16) talked about the process of making a decision balancing different tensions. This refers to an informed decision-making process, usually taken after thorough reasoning and the balancing of different factors. It is an examined act after the process of weighing the importance, value, practical impact of personal interest, and/or other factors. Students also discussed the uncertainty/limitations/implications that are associated with making decisions.

These decisions can be about the students' research projects; here is a quote from Ron:

> *In the end, I still chose the ones that I liked, instead of choosing one (project) because it can produce papers easily, or, say doing a project because the boss likes it. I think, I am doing what I like now. Yeah.*

-Ron

Codes:

Commitment - Demonstration - To make a decision balancing different tensions

Commitment - Demonstration - Stylistic attributes (to do research out of real interest vs. to do research for the sake of doing research)

The decisions also include one's career choices:

> *E: Um, it's about my plan for future; I decided to not stay in university as a teacher or as a professor.*
>
> *R: Can you elaborate, how did you make this decision?*
>
> *E: Um, it's, related to, it's related to my interest. I hope that I can do some practical things. Moreover, it's also because of the impact from my environment, impact from the people I see, it is quite demanding to be a professor. It requires much work, work with long hours. Also, you will have to think about your research project every day. So I think, for myself, from my perspective, the decision is made for my own benefit; meanwhile it's influenced by other factors, environment, people or things around me.*

-Ray

Codes:

Commitment - Demonstration - To make a decision balancing different tensions

Cameron described the process of making choices as:

> *In the beginning, you know that there is such a way out there, but you are not very sure, in the process of pondering about it, it's like, the degree of certainty goes up a little bit, it keeps going until a certain threshold, then you make a decision…It is basically a process*

of, say, in the beginning, you choose it only because you really want to. But your, your reasoning tells you, "You are not very sure yet." But, then, if you have the component of pure emotion going down, then, it can kind of balance with the rational component, then you can make a decision.

-Cameron

Code:

Commitment - Demonstration - To make a decision balancing different tensions

Here is another quote about making choices from Ryan. Interestingly, he also spoke of the possible limitations of making choices or even reasoning itself. In addition, he seemed to have learned to accept the limitations of making choices and the consequences related with these limitations.

You make choices. Of course, the reason why you cannot regret after making choices is that, before making a choice, in fact you have considered all the factors. You made the choice in a very serious manner. So you cannot regret. If you made the decision in a hasty manner, in the end, the chance for regretting is huge. If you have considered all kinds of factors in a holistic manner, in that case, right, because the situations can change, you made the choice, after that, maybe it did not turn out the way you wanted, maybe you will have some regrets. Um, but, first, there is no use even if you regret about it. Therefore, your understanding, should, your thinking should be changed, you should not regret. Actually, you can adjust yourself. Because people regret, it is natural. If you did not reach the result you planned, you wanted, you may regret about the decision. You may think, "I lost a lot of things." But, this is something that can be adjusted. Little by little, from a certain thing that you regret about, you adjust yourself, you do that intentionally. After a while, that becomes a habit, then you won't regret. This is very important.

-Ryan

Codes:

Commitment - Demonstration - To understand the uncertainty/limitations/implications of making decisions

Commitment - Demonstration - To experience, and/or understand and/or accept the limitation of reasoning

Commitment - Demonstration - To make a decision balancing different tensions

Commitment - Demonstration - Stylistic attributes (being prudent vs. being impulsive)

Life Planning

Another major demonstration of commitment to relativistic thinking is the different aspects of life planning in which students were involved. Students spoke of the significance of having a clear goal (8 out of 16), life planning (4 out of 16), the active preparation for their careers (4 out of 16) and 19, and to recognize the need to prioritize (2 out of 16). Here is an example from Charles:

Then, it's about my career planning, what am I going to do after graduation? I struggled for a long time for this decision. But, this decision, I feel, I did not follow others blindly. I made the decision according to my thoughts and my entire goal of life. I went abroad later, you know, but I faced a lot struggles at that time. Because, you know that, there are other options (in China), say I can go to the graduate school in my university without entrance exams, I can go to other good universities in China without entrance exams, or there were many other good offers. But I made the decision to come to the US; it was after a long time of struggling, because I gave up some other things, I balanced different things for a long time.

-Charles

Codes:

Commitment - Demonstration - To do life planning

Commitment - Demonstration - To have a clear goal

Commitment - Demonstration - To make a decision balancing different tensions

And here is what Ray said about his goal:

In fact, because, after finished undergraduate studies, we (Chinese students) decided to go broad, we were coming with our dreams, with the hope to start some, to do something that we really wanted to, to learn the things we really wanted to learn.

-Ray

Code:

Commitment - Demonstration - To have a clear goal

Stylistic Issues

Many decisions are made in a concrete area, such as one's career, research project, and so on. However, there are many other decisions that are made in a more stylistic manner according to Perry, that is, the students "seemed to express the qualitative experience of Commitment in polarities" (1970, p.161). The Chinese engineering doctoral students also exhibited these stylistic attributes, such as to do practical research versus to do basic research, to do research out of real interest versus to do research for the sake of doing research, and/or to live a rich life versus to live by essentials, etc.

It should be noted that codes like these may look like a judgment call is being made as to which of the two polarities is better; however, the focus here is NOT on the results of the judgment call. Instead, the focus stands in relation to the students' thinking and reasoning processes, those of which lead to them making essential judgment calls for their own lives. That is to say, the emphasis here is that the interviewee went through a complicated reasoning process that allowed him/her to make a decision. And the decision the interviewee made usually involves some stylistic attributes, meaning that the individual has decided to commit to a style of doing things or living his/her own life.

Here are some examples:

…Because I want to produce something that other people can use. I don't want that, after three years, I produced something, had published papers on it, but even I myself don't want to use it. I would feel that I am wasting my time and energy. So I think, if I go to industry, even a small change can become a product. And even a small change to the world, I would feel my life is meaningful.

-Rebekah

Codes:

Commitment - Demonstration - Stylistic attributes (to do practical research vs. to do basic research)

And:

In the beginning you need to use your willpower, you cannot hope to, you should realize that it requires willpower. But the purpose for using your willpower is to help you go deep into the subject. After you go deep into the subject, you will find the fun of doing it. But if you never become interested in the things you are learning, you are only interested in getting a job after learning it, you are only interested in getting the course credit, and you never get interested in the materials themselves, then you will always have to use your willpower, in that case you will never be good at it.

-Rose

Code:

Commitment - Demonstration - Stylistic attributes (learning/doing research out of real interest vs. learning/doing research for the sake of doing it)

In addition to the stylistic attributes shown within a research/study, some stylistic attributes are demonstrated in other aspects of life. Here is a quote drawn from Rose to illustrate this point:

So I started to cultivate some passions for life. Later I think, I should not, learning and research are things that you have to do. But, in addition to these things, you need passions for life, in that way, you will live a rich life, otherwise you will have no inspiration. So now I take up a lot of hobbies...Before, I never engaged myself in these. Life has decreased to the essentials, only about the things that I have to do. Therefore I think, it's like, it's important to cultivate interest. It will help you stay active in your spirit. It keeps you from being negative; it inspires you in your research and learning.

-Rose

Code:

Commitment - Demonstration - Stylistic attributes (to live a rich life vs. to live by essentials)

Attributes

Along with the outer exhibitions of commitment in one's career and life planning, the active undertaking of responsibilities and the informed decision-making in concrete areas and in a stylistic manner, many students talked about those attributes that they feel are needed in the process of commitment. Perry described that the central theme for development toward commitment is responsibility. From the interviews of the Chinese engineering doctoral students, it seemed that the under-taking of major responsibilities is accompanied with the exhibitions of different attributes, such as being self-motivated, being persistent despite uncertainty, having a positive attitude, and/or a strong motivation and been devoted to one's work. Here is an example drawn from Cody:

Uh, what helped me, actually, sometimes, you know, my human potential, I mean, the potential is actually activated by some pressures. For example, I want to graduate, I have to do it by that certain point, otherwise, I won't get the degree down, so, I have to do it. There is no other choice. Sometimes I asked, actually, you know, I need some materials, for the experiments, so I need to, I go to companies and ask for their donations, you know... because I really need them. So, it is really, I have to do it. Even you know, even if I got rejected, it is no big deal, just move on, go to another one, asking for donation.

-Cody

Codes:

Commitment - Demonstration - To be persistent despite uncertainty

Relativistic Thinking - Factor - Failures, pressures, or obstacles

A positive attitude can help students in facing the diversity of opinions and critiques from others:

First thing I learned was to be a person "full of positive energy," and a person "full of positive energy," will, take a lot things, like I said, take the critiques from others, or other things done to you, consider these things in the right direction. By "positive energy," I mean, it's not pleasing the people around you, it's really, whatever problem you encounter, you strive toward the positive direction. Maybe you can be down sometimes, but you make efforts.

-Rebekah

Codes:

Commitment - Demonstration - To be persistent despite uncertainty

Commitment - Demonstration - To have a positive attitude

Inner Strength/Faith

Amid the complexity of committing to relativistic thinking as a way of life, students sensed the limit of reasoning and constraints during making a decision even after careful reasoning. According to Perry's finding, one student realized that he "had to work from a point of faith" (Perry 1970, p. 184). Some students here also mentioned inner strength or faith as something they hold on to.

For example, here is what Cody said about faith:

E: I think it really depends on how you behave, I mean, your attitude, stay positive, just believe in yourself, I can do it, even if it is new to me.

R: So you are talking about confidence.

E: I will say faith, yes, faith.

R: You mean faith in yourself?

E: Yes, faith in myself, I believe as long as I put in effort, I will do it; I can make it.

-Cody

Code:

Commitment - Demonstration - To commit to a certain type of faith

The elements of faith to which these students have referred include faith in one's inner strengths or faith in some types of religion. The students were quite willing to discuss how they felt about transforming their ways of lives because of their faith:

Um, it (Christian faith) changed me a lot, my attitude towards life, towards my work, things, people around me, and some, and my whole worldview.

-Rena

Code:

Commitment - Demonstration - To transform one's way of life because of his/her faith

9.4   Discussions

To summarize, the findings based upon qualitative data confirm and refine the profiles gathered from the survey results through the operationalization of the four different thinking styles of Perry's theory, namely, dualistic thinking, multiplistic thinking, relativistic thinking, and commitment to relativistic thinking, and through the distributions of the frequency counts of participants in the instances of demonstration for each thinking style. The participants that are categorized in the five different groups did represent distinctive distributions of the frequency counts among the four thinking styles.

In addition, the operationalization of the four different thinking styles provides practical instances of these thinking styles involved in research, learning, and other areas for these Chinese engineering doctoral students. These instances and their frequencies of occurrences among the different groups provide further details for the students' epistemological development.

Specifically, corresponding to the survey results, which indicated that most of the students' prominent thinking styles fell within the higher levels of thinking (i.e., *Relativism* or *Commitment*), the main findings based upon the qualitative data were then derived from the participants from the groups of *Relativism, Relativism-Commitment,* and *Commitment.* The main findings and the connections of these findings to the current literature can be summarized in the following aspects

Operationalization of Relativistic Thinking

First of all, the demonstrations of relativistic thinking among Chinese engineering doctoral students are shown in different areas. In their interactions with their advisors or professors, on the one hand, these students were able to learn from their professors/advisors' ideas or philosophies, while, on the other hand, they strived to learn/think/conduct research in an independent manner. In their interactions with other students or researchers, these students came to appreciate and actively seek out different feedbacks and even enjoyed the process of being critiqued, while they also reserved the right to discern and test the diverse feedback and opinions for themselves. In addition to these traits, the students strove to produce sound research results and explore and test the unknown while trying to master knowledge at a deep level.

Prior findings from Jiang (2010) about Chinese engineering doctoral students suggested their development of independent thinking and learning skills for these students. The students regarded their advisors as guides, but similarly realized that it was their responsibility to find and solve problems. As found from the survey results, a major portion of students (nearly 80 %) fell within the higher levels of thinking (i.e., *Relativism* or *Commitment*) in the context of Perry's theory. The skills and traits of students acknowledged in Jiang's study were identified here as part of the demonstrations of the higher levels of thinking. These demonstrations include skills or abilities such as learning/thinking/conducting research in an independent manner, finding and/or solving problems, and so on. Similar to Jiang's report, students within this study appreciated the deep thinking and/or philosophy of their professors/advisors and considered them as a guide in their study/research. Nevertheless, students were also able to reserve the right to discern and test feedback or information, which is the essence of a relativistic thinking described in a number of studies that followed the line of Perry's theory (Kuhn 1991; Baxter Magolda 1992; King and Kitchener 1994).

Operationalization of Commitment to Relativistic Thinking

With the prominent demonstrations of relativistic thinking, many of those students classified within the *Relativism* group have started to show demonstrations of commitment to relativistic thinking. These demonstrations are reflected in their attempts to assume major responsibilities, take leadership in their research/ learning/other areas of life, and make informed decisions while accepting the limitations or implications of these decisions. In addition to the decision-making involved in some concrete life areas (such as career and study), they have also made some other decisions in a stylistic manner, which are often expressed in polarities.

Concerning the demonstrations of the style of commitment to relativistic thinking, the stage of *Commitment* is the least well-defined stage among Perry's four major stages (Moore 2002). Moreover, among the five theoretical frameworks in Perry's theory (as described in Sects. 2.2–2.4), this stage is only described in Perry's original intellectual and ethical development model. The stage of commitment was not further explored in the later theoretical frameworks along Perry's theory. Therefore, what is known about this stage can only be derived from Perry's original theory. According to Perry's description, responsibility is the major theme for the stage of commitment:

> The drama of development now centers on this theme of responsibility. The hero makes his first definition of himself by some engagement undertaken at this own risk... as he expands the arc of his engagements and pushes forward in the impingements and unfoldings of experience, he discovers that he has undertaken not a finite set of decisions but a way of life. (Perry 1970, p. 170)

Similar to Perry's descriptions, the students who are classified in the groups of *Relativism, Relativism-Commitment,* and *Commitment* started to take on the major responsibilities in research/learning and strove to be the lead of themselves in different areas. Also, similar to Perry's findings about the interesting polarities in students' decision-making, the students involved in this study demonstrated some stylistic attributes in their choices made. All of these findings not only confirm the existence of the commitment to relativistic thinking among Chinese engineering doctoral students, but also provide practical instances for reflections regarding the commitment to relativistic thinking among Chinese engineering doctoral students.

9.5 Conclusion

With findings based upon qualitative data and the analysis, this study confirmed and refined the profiles gathered from the survey results through the operationalization of the four different thinking styles of Perry's theory, namely, dualistic thinking, multiplistic thinking, relativistic thinking, and commitment to relativistic thinking. The operationalization of the four thinking styles within the Perry's theory offers practical instances into these thinking styles. The distributions of the frequency counts of participants in the instances of demonstration for each thinking style validated their groupings obtained from the survey results.

Using qualitative data, this study expanded the current understanding of Chinese engineering doctoral students' epistemological development by the operationalization of relativistic thinking and commitment to relativistic thinking in particular. The practical instances identified in the qualitative data, especially the instances of relativistic thinking and commitment to relativistic thinking, can potentially be used as a foundation for establishing engineering-specific assessment tools for the measurement of higher levels of thinking (i.e., relativistic thinking and commitment to relativistic thinking) in the context of Perry's theory.

References

Baxter Magolda, M. B. (1992). *Knowing and reasoning in college*. San Francisco: Jossey-Bass.

Belenky, M. F., Clinchy, B. M., Goldberger, N. R., & Tarule, J. M. (1986). *Women's ways of knowing: the development of self, voice and mind*. New York: Basic Books.

Cicchetti, D. V. (1994). Guidelines, criteria, and rules of thumb for evaluating normed and standardized assessment instruments in psychology. *Psychological Assessment, 6*, 284–290.

Cohen, J. (1960). A coefficient of agreement for nominal scales. *Educational and Psychosocial Measurement, 20*, 37–46.

Creswell, J. W. (2008). *Research design*. London: Sage Publications.

Gwet, K. L. (2012). *Handbook of inter-rater reliability: The definitive guide to measuring the extent of agreement among multiple raters*. Advanced Analytics, LLC.

Hruschka, D., Schwartz, D., St. John, D., Picone-Decaro, E., Jenkins, R., & Carey, J. (2004). Reliability in coding open-ended data: Lessons learned from HIV behavioral research. *Field Methods, 16*, 307–331.

Jiang, X. (2010). *Chinese engineering students' cross-cultural adaptation in graduate school*. ProQuest: The University of Michigan.

King, P. M., & Kitchener, K. S. (1994). *Developing reflective judgment: Understanding and promoting intellectual growth and critical thinking in adolescents and adults.* San Francisco: Jossey-Bass.

Kuhn, D. (1991). *The skills of argument.* Cambridge: Cambridge University Press.

Landis, J. R., & Koch, G. G. (1977). The measurement of observer agreement for categorical data. *Biometrics, 33,* 159–174.

MacQueen, K. M., McLellan, E., Kay, K., & Milstein, B. (1998). Codebook development for team-based qualitative analysis. *Cultural Anthropology Methods, 10,* 31–36.

Merriam, S. B. (2002). Introduction to qualitative research. In S. B. Merriam (Ed.), *Qualitative research in practice.* San Francisco: Jossey-Bass.

Merriam, S. B. (2006). Transformational learning and HIV-positive young adults. In V. A. Anfara Jr. & N. T. Mertz (Eds.), *Theoretical frameworks in qualitative research* (pp. 23–38). Thousand Oaks: SAGE Publications.

Miles, M. B., & Huberman, A. M. (1994). *Qualitative data analysis.* Thousand Oaks: Sage.

Moore, W. S. (2002). Understanding learning in a postmodern world: reconsidering the Perry scheme of intellectual and ethical development. In B. Hofer & P. Pintrich (Eds.), *Personal epistemology: The psychology of beliefs about knowledge and knowing.* Mahwah: Lawrence Erlbaum Associates.

Patton, M. Q. (2002). *Qualitative research & evaluation methods.* Thousand Oaks, CA: Sage

Perry, W. G. (1970). *Forms of intellectual and ethical development in the college years: A scheme.* New York: Holt, Rinehart and Winston.

Wang, W. (2009). *Chinese international students' cross-cultural adjustment in the U.S.: The roles of acculturation strategies, self-construals, perceived cultural distance, and english self-confidence.* Austin: The University of Texas. Retrieved from ProQuest.

Ye, Y. (1992). *Chinese students' needs and adjustment problems in a U.S. university.* Lincoln: The University of Nebraska. Retrieved from ProQuest.

Chapter 10
Factors Associated with Advanced Epistemological Thinking

This chapter focuses on confirming and refining the results from the quantitative data analysis by investigating the factors that are related to the overall profile presented in Chap. 7 in which the prominent epistemological thinking style of nearly half of the students was relativistic thinking. In this chapter, factors that were associated with students' epistemological development to advanced thinking were explored. These factors include impacts from personal aspects, experiential influences, and contextual dimensions.

As mentioned in the Sect. 9.2 Methods section, it is very difficult conceptually to separate the factors that are associated with relativistic thinking and the factors that are associated with a commitment to relativistic thinking. Although the factors were named as Relativistic Thinking-Factors, they also contribute to the commitment to relativistic thinking because the term "commitment" actually means a form of commitment to relativistic thinking. Therefore, Relativistic Thinking-Factor codes are used to indicate the factors that are related either to relativistic thinking or to commitment. The factors mentioned here are again gathered from the groups of *Relativism, Relativism-Commitment,* and *Commitment.*

It should be noted that the factors identified here should not be isolated and regarded only as the reasons that facilitate relativistic thinking. That is to say, while these aspects are classified as factors in this section should in no way prevent them from serving as reasons intended to promote other styles of thinking, such as the style of multiplistic thinking. Essentially, these factors can probably facilitate other types of thinking, too. They are defined within this section of the research text only because they are identified as associated with relativistic thinking. On the other hand, the factors listed here are by no means an exclusive list of the many possible factors that can facilitate the development toward relativistic thinking.

Potentially some overlaps exist between the concepts of "demonstrations" and "factors." The term "demonstrations" refers to current or past activities or behavioral patterns that reflect the characteristics of the thinking or a direct expression of the particular thinking process. The term "factors" may include some descriptions of experiences, but often involves the reasoning process concerning how certain

© Springer Science+Business Media Singapore and Higher Education Press 2017 145
J. Zhu, *Understanding Chinese Engineering Doctoral Students in U.S. Institutions,*
East-West Crosscurrents in Higher Education, DOI 10.1007/978-981-10-1136-8_10

experiences or other components contribute to the students' thinking process. It should be noted that many quotes examined in this research were regarded as having the elements of both the demonstration and the factor.

The total number of the Relativistic thinking-Factor third level codes is 45. The Relativistic Thinking-Factor codes with counts no less than 5 across all three groups are shown in Table 10.1. A complete list of all Relativistic Thinking-Factor codes can be found in Appendix I. These factors are grouped into three categories (Table 10.2).

10.1 People Factors

The roles and functions of advisors or course instructors are prominent in helping the students as they develop relativistic thinking. The factor code "Advisors" ranks highest among all of the factor codes, the code of "Senior lab mates or school mates" ranks third, and the "Course instructors" code ranks fourth. Some students (4 out of 16) also talked about the role of their parents in their thinking development.

Here is an example from Rebekah in which she spoke about the role of her advisor:

> How did I realize these (ideas)? It is when you encountered a bottleneck, in fact, you are striving, that is, by "striving," in my boss' word, actually I asked him, I think my boss is a person with very strong philosophy, I was asking him, I said, "How did you come up with that idea?" He said, "Just try! Try, and try, then it comes out." Actually, what I am saying to you now, is what he said to me. In the beginning, I wasn't very clear about it, but then later, I got it. That is, it's fine, you just, little by little, that will be some time, you find yourself stuck, then you will start to find a solution. That is, you finally find that, there is no door in the room, then you start to think, should I climb to the window, or even climb to the ceiling? Therefore, I think, you get to the desperate point; naturally, the nature to survive is kindled. Then you will some more, yeah.

—Rebekah

Codes:

Relativistic Thinking—Demonstration—To regard professors/advisors' role as a guide

Relativistic Thinking—Demonstration—To appreciate the deeper thinking/philosophy of professors/advisor

Relativistic Thinking—Factor—Advisor

Relativistic Thinking—Factor—Failures, pressures, or obstacles

Commitment—Demonstration—To be brave/to overcome challenges

The role of course instructors is also important to many other students; for example, here is what Rick had to say:

Table 10.1 Relativistic thinking-factor code list for the groups of *relativism, relativism-commitment,* and *commitment*

Third level codes	Rick	Robert	Ron	Rena	Ryan	Ruby	Rose	Rebekah	Ray	Ken	Kirk	Kevin	Cameron	Cody	Charles	Charlie	Total	Count of students
1. Advisors	1	0	3	3	3	3	0	4	2	3	1	1	3	0	4	0	31	12
2. Discussions with other students	2	1	3	0	0	5	3	1	0	3	2	2	1	2	4	1	30	13
3. Senior lab mates or school mates	3	2	4	1	3	3	1	2	0	1	1	0	1	0	4	1	27	13
4. Course instructors	2	0	0	1	0	2	3	0	0	1	2	4	0	1	2	3	21	10
5. Failures, pressures, or obstacles	1	5	0	1	1	1	0	4	0	0	0	1	0	2	0	1	17	9
6. Exposures to course work in the US	2	1	2	0	2	2	1	0	0	0	2	2	0	0	0	0	14	8
7. Being in the US	0	1	4	0	1	0	0	1	0	0	2	2	0	0	2	0	13	7
8. Experiences with group projects	2	1	1	1	0	0	2	0	0	0	2	2	0	0	0	0	11	7
9. Positive or negative role models	0	1	0	1	0	2	3	1	1	0	0	0	1	0	1	0	11	8
10. Being in a US graduate school or university	1	1	0	0	0	1	0	1	0	0	2	1	1	0	2	0	10	8
11. Research group	1	0	5	0	1	0	0	0	0	1	0	0	2	0	0	0	10	5
12. Encounters with open-ended problems/projects	1	1	2	2	1	0	0	0	0	0	0	1	0	0	0	0	8	6
13. Parents	0	0	0	1	0	0	2	0	0	2	1	0	0	0	0	0	6	4
14. Professional conferences	0	0	1	1	0	0	0	0	0	0	2	0	0	1	1	0	6	5

(continued)

Table 10.1 (continued)

Third level codes	Rick	Robert	Ron	Rena	Ryan	Ruby	Rose	Rebekah	Ray	Ken	Kirk	Kevin	Cameron	Cody	Charles	Charlie	Total	Count of students
15. Experiences of studying in a group	0	1	0	0	0	0	0	0	0	0	0	0	0	3	1	0	5	3
16. Experiences with oral presentations	0	1	0	0	0	0	0	0	0	1	1	1	1	0	0	0	5	5
17. Learning to collect and/or read academic papers	0	0	1	0	0	2	1	0	0	0	0	0	1	0	0	0	5	4

Table 10.2 Three main categories for relativistic thinking-factors codes	*People factors*
	• Advisor
	• Senior lab mates or school mates
	• Course instructor
	• Positive or negative role models
	• Parents
	Experiential factors
	• Discussions with other students
	• Failures, pressures, or obstacles
	• Exposures to course work in the US
	• Experiences with group projects
	• Encounters with open-ended problems/projects
	• Experiences of studying in a group
	• Experiences with oral presentations
	• Learning to collect and/or read academic papers
	Contextual factors
	• Being in the US
	• Research group
	• Being in a US university or graduate school
	• Professional conferences

Concerning the stuff in my field, I learned quite a lot of things. Because, first of all, um, first, on one hand, from the perspective of taking classes. Because the professors here (in the US), their classes have, how to say, they have their own styles. Even for, say similar content to what I learned in China, they usually won't follow the textbook. They put in a lot of their own thoughts from their research, and some new thoughts into the courses. Just from the courses, I can learn many things that I cannot learn from being in China. Because, here, I mean, the styles are different.

—Rick

Codes:

Relativistic Thinking—Factor—Course instructor

Relativistic Thinking—Factor—Exposure to the course work in the US

And the roles of senior lab mates or classmates were often mentioned among students, too. These lab mates or classmates were "senior" in the sense that they helped the interviewees in certain aspects. It does not mean that they were always senior in their academic years. Here is an example from Robert:

...It is important, I mean, don't look down upon anybody. I think this is important. Even, some undergraduates...maybe, my thoughts are, my abilities now, are probably not as good as some undergraduates, in many ways, I am not as good as they are, I think, I should hear their thoughts. It's not like, you are older than them, you are senior than them, then you can think what they say are meaningless, it's definitely not the case. I feel that many classmates, what they learned are some things that I just heard about. I think I should learn more from them. Even when you are discussing the same problem, maybe you yourself think you got everything figured out. But when you are talking to others, you find that he has come up with something that you never thought about. ...When we are doing the homework, some of the things talked about by several students, maybe even the instructor had never thought about those things when he put the problem set together, he never thought that it could be considered in that particular way. Yeah.

—Robert

Codes:

Relativistic Thinking—Factor—Senior lab mates or school mates

Relativistic Thinking—Demonstration—To actively ask others questions or seek out help

Relativistic Thinking—Demonstration—To appreciate conflicts and/or the diversity of opinions

Relativistic Thinking—Demonstration—To communicate with people from diverse backgrounds

And several students talked about the role of parents in shaping their thinking:

Um, my parents they often allow me to decide myself. They never came up with ideas for me, to ask me to follow the path they designed for me. Instead, they let me think for myself, to find what I wanted to do. It helped me to have the habit of independent thinking, say, not relying on others, but to think for myself.

—Rena

Codes:

Relativistic Thinking—Demonstration—To think in an independent manner

Relativistic Thinking—Factor—Parents

10.2   Experiential Factors

Different experiences that are potentially associated with the development of relativistic thinking were frequently mentioned among these students. These experiences include discussions with other students (13 out of 16), failures, pressures, or obstacles (9 out of 16), exposures to course work in the US (8 out of 16), experiences with group projects (7 out of 16), experiences with open-ended problems or projects (6 out of 16), and so on.

As has been indicated in earlier quotes, having discussions with other students seemed to be very prevalent among these studies participants. The exposure to a diversity of opinions seemed to help these students in the development of their thinking:

In our group, sometimes, some American students, they have some opinions, some ideas that I have never thought about. For example, when designing a part, I may only think about its functions, but he (they) would think from a bigger scope. Say, agronomics, that is, the interaction process between human and machines. He (they) would even think about the impact of the product to the environment, which I would never think about. I only think about whether the user feels comfortable with the product. But they would think from different perspectives, to make you the user, instead of the designer.

—Ron

Codes:

Relativistic Thinking—Factor—Discussions with other students

Relativistic Thinking—Factor—Students from other countries/backgrounds

Relativistic Thinking—Factor—Senior lab mates or school mates

Relativistic Thinking—Demonstration—To appreciate conflicts and/or the diversity of opinions

Relativistic Thinking—Demonstration—To communicate with people from diverse backgrounds

And the role of failures, pressures, or obstacles seems to be closely related with the students' development of thinking. The following quote is part of a quote cited before, intended to emphasize this fact:

…When you come to a point, you find, there is no way, then you, probably you will start to find a way out. You finally realize that, there is no door in the room, then you start to think, should I climb to the window, or climb to the ceiling? Therefore, I think, you get to the desperate point; naturally, the nature to survive is awakened. Then you will try some more, yeah.

—Rebekah

Codes:

Relativistic Thinking—Factor—Failures, pressures, or obstacles

Commitment—Demonstration—To be brave/to overcome challenges

Here is an example related to the factor of being exposed to course work in the US:

Because in the US, you can pick whatever classes you want; before, in China, I didn't have many choices, just to go to this one or that one, because I only have these many choices. But here in the US, you can pick all the courses from the university. Um, so I would say, the difference is that you can learn more things, to get to know more things. Um, right, because the research direction we are in, it is not a narrow direction, because it is related with many areas, it relates to mathematics, physics, um, so, it relates to computer sciences, so, it's interdisciplinary, so, you can take more classes, learn more things.

—Kirk

Codes:

Relativistic Thinking—Factor—Exposure to the course work in the US

Relativistic Thinking—Factor—Being in the US

Relativistic Thinking—Factor—Being in a US graduate school or university

Relativistic Thinking—Demonstration—To expand one's scope of knowledge (with the intention to inform one's research/study)

The experience of group projects, either recommended or required by a professor, also seemed to be helpful to many students, although not every student was comfortable with it:

> Actually even till now I am still not very comfortable with this method, because I, how to say, I prefer to be by myself, to read by myself, to understand by myself. I mean, now, because after I came here, many things, you have to force yourself to change. For example, the professors assigned to you tasks, say, you guys in this group write a paper together, or do a presentation together. In many cases, it is imposed, that is, we have to form this group. Many times, there are different people from different cultures in the group that the professor assigned. Then, you have to force yourself to accept, the things that are different from your culture.

—Rick

Codes:

Relativistic Thinking—Factor—Experiences with group projects

Relativistic Thinking—Factor—Exposure to the course work in the US

Relativistic Thinking—Factor—Course instructor

Relativistic Thinking—Factor—Being in a US graduate school or university

Relativistic Thinking—Factor—Students from other countries/backgrounds

10.3   Contextual Factors

Some students discussed the importance of the environment around them. These different contexts include being in the US, being in a US university or graduate school, being in a research group, and being in a professional conference.

Here is an example from Ron about his general feeling regarding his opportunity to be in the US:

> Compared to the US, China is a relatively laid-back environment. So in a laid-back environment, I said that people can be lazy, so you may play PC games, you do a lot of sports, you may want to be involved in a lot of entertainment. But, here, the environment is, the way other people are may give you, like, you will have a sense of competitiveness. You will need to imagine, you will need to imagine, what it is like to collaborate with someone here, what it is like to argue with him. It is a dynamic process. You learned a lot through it, you learned to shape your character.

—Ron

Codes:

Relativistic Thinking—Factor—Being in the US

Relativistic Thinking—Factor—Being in a competitive environment

While some student finds that it is helpful to be in a competitive environment, other student may prefer the opportunities they have to work within a collaborative research group:

...He (the advisor) will ask our designs, our thoughts, and then he will help revise a little bit. Also, he encourages us, to communicate among the group, to help each other. I really like this environment. But before, my former research group, what my former advisor liked was kind of a competitive environment. I guess that is not suitable for me (laugh).

—Ken

Codes:

Relativistic Thinking—Factor—Research group

Relativistic Thinking—Factor—Advisor

Relativistic Thinking—Factor—Discussions with other students

Several students find that it is useful to attend professional conferences:

Okay, alright. What I learned, I think, from my standpoint, it's always important to keep yourself social, I mean, not just in your daily life, but also in academia. By talking to people, so you know what is going on in your field of other people, what they are doing. If I go to some conference, you know, the conference is a good way to mingle, to socialize and mingle with the people from the same field, learn what is their research,... What I, how we can learn from each other, so I will say, open your mind, to, just to talk with people, socializing, these things, that will help each other.

—Cody

Codes:

Relativistic Thinking—Factor—Professional conferences

Relativistic Thinking—Demonstration—To be aware of what is going on in the field (with an explicit intent to inform one's own research or study)

Commitment—Demonstration—Stylistic attributes (to be social professionally vs. to be isolated)

10.4 Discussion

In summary, different factors, such as people, experiences or contexts that are related to the students' relativistic thinking or commitment to relativistic thinking are summarized in this chapter. The top factors that were identified among the students to be associated with relativistic thinking and commitment to relativistic thinking include the elements of people around them, the contexts in which they were exposed, and/or those experiences with which they were involved. Students emphasized the impact from their advisors, course instructors, and lab mates or schoolmates. Many students spoke of their personal experiences where they discussed with students who had helped them with their thinking. Also, it seems like the students appeared to realize that their opportunities to experience failures, pressures, or obstacles also simulated their thinking.

Some of the factors identified here are similar to findings in the current literature. For example, the importance of advisors in students' development of thinking was stressed in Jiang's study (2010). Also, the factor of being in the US and the related freedom of choices in research and course work were also mentioned in Jiang's study. Another of the factors identified here was that of the students' encounters with open-ended problems/projects. Pavelich and Moore highlighted the importance of experiential learning through the use of complex, real-world problems in helping students' development toward more mature thinking in Perry's model (1996). Here, these factors have been identified to be associated with relativistic thinking or the commitment to relativistic thinking in the context of Perry's theory. The different factors identified here along the demonstrations of relativistic thinking and the commitment to relativistic thinking together constitute a more complete picture about the students' epistemological development and potentially provide practical implications for teaching and learning.

10.5   Conclusion

In conclusion, the investigation of the factors that were associated with relativistic thinking and commitment to relativistic thinking among the studied participants constitutes a more holistic picture of students' epistemological development. The factors that were identified as associated with higher levels of thinking in this study can possibly offer guidance to current educational practices to facilitate the development toward more advanced and complex manners of thinking among engineering students.

References

Jiang, X. (2010). *Chinese engineering students' cross-cultural adaptation in graduate school.* Retrieved from ProQuest: The University of Michigan.
Pavelich, M. J., & Moore, W. S. (1996). Measuring the effect of experiential education using the Perry model. *Journal of Engineering Education, 85*(4), 287–292.

Chapter 11
Lessons Learned and Future Directions

Based upon the findings obtained through both the quantitative and qualitative data of this work, this chapter first situates these major findings within the current literature and presents the main lessons learned from this research. The limitations and potential ways to extend this research are then discussed in greater detail. Future directions are further elaborated regarding the possible ways to expand this work to provide a practical impact on students' epistemological development.

11.1 Lessons Learned

This research first provided a survey that reflects all of the four developmental stages of Perry's theory. This work then utilized an explanatory research design, in which quantitative survey results provided an overall trend of the Chinese engineering doctoral students' epistemological developmental stages and results based upon qualitative data, which largely confirmed the findings from the survey results. Moreover, the results derived from the qualitative data confirmed and refined the overall trend of the survey results by exploring the practical instances of each epistemic thinking style. By combining the findings based upon both quantitative and qualitative data, this work provided a more holistic picture of Chinese engineering doctoral students' epistemological development.

First, the findings from both the quantitative and qualitative data seem to suggest that many of the Chinese engineering doctoral students demonstrated relativistic thinking, which is probably related to their doctoral educational experiences. These experiences include their engagement in a research team, their exposure to US graduate course work, their interactions with their advisors/lab mates/classmates, and their research experiences; said research experiences include their opportunities for completing a literature review, designing and conducting experiments, presenting their work, and so on. The prevalent presence of these instances suggests

© Springer Science+Business Media Singapore and Higher Education Press 2017
J. Zhu, *Understanding Chinese Engineering Doctoral Students in U.S. Institutions*,
East-West Crosscurrents in Higher Education, DOI 10.1007/978-981-10-1136-8_11

the close relationship existent between the US doctoral education and these students' development of relativistic thinking.

The presence of many of these practical instances reflects the training goals of doctoral programs. The desired characteristics for PhD recipients discussed among current literature include the abilities to conduct research in an independent manner (Gardner 2008), the skills to think in a critical manner (Cox et al. 2011), the abilities to explore the new and unknown areas of their disciplines (Golde and Walker 2006), and so on. Many of these skills or abilities also represent the core ideas of relativistic thinking and commitment to relativistic thinking. For example, one of the identified demonstrations of relativistic thinking is to produce sound research results. Steps taken to produce sound research results imply an analytical thinking process in which a student examines one's own research in a critical manner, tests and evaluates the research results, and possibly supports the results with evidence, many of which are the key ideas of relativistic thinking. Meanwhile, this whole process of producing sound research results is also closely related to the training goals of the doctoral education.

Second, by focusing on students' epistemological development through use of a quantitative method to provide an overall picture and a subsequent qualitative method to explore in-depth stories, this work provides a more complete picture of Chinese engineering doctoral students' epistemological developmental status. On one hand, the main findings from the survey results illustrated that nearly 80 % of the students showed a prominent thinking style in relativistic thinking and/or commitment to relativistic thinking. The prevailing presence of higher level thinking, i.e., relativistic thinking and commitment to relativistic thinking in the findings, contrasts sharply with those prior findings by Zhang and colleagues, which suggested that Chinese college students in China demonstrated a developmental trend that was in opposition to the trend described in Perry's theory (Zhang 1995, 1999, 2000, 2002; Zhang and Hood 1998; Zhang and Watkins 2001). It is reasonable to speculate that the main difference between this current research and the prior findings resides primarily in the differences of the research populations. That is, the population in this research, i.e., *the Chinese engineering doctoral students in US programs*, probably differs to a great extent from the population involved in the earlier studies by Zhang et al. in which the focus was on the Chinese college students within China (Zhang 1995, 1999, 2000, 2002; Zhang and Hood 1998; Zhang and Watkins 2001).

On the other hand, the findings of this current work, especially with regard to some of the instances identified in the qualitative data, do seem to resemble some of the findings existent in Jiang's research (2010). For example, she identified the Chinese engineering doctoral students' efforts in learning to be independent researchers. Nonetheless, Jiang's focus was on these Chinese engineering doctoral students' adjustment to the US graduate programs. Therefore, he did not provide further instances that were related to students' thinking. With a focus on the exploration of the demonstrations of different thinking styles, this current work has provided a number of practical instances as to Chinese engineering doctoral students' thinking in the context of an engineering doctoral program.

Third, the results also indicate a larger context in addition to the specific context of engineering doctoral programs as related to students' epistemological thinking; these factors may be further associated with the development of different styles of epistemological thinking. For example, some of the students discussed how the factor of being in the US had shaped their way of thinking about work and life. Some other students spoke about the impact of their parents on their development of thinking while they grew up. The existence of these varying factors has further extended the scope of the current research into a larger sociocultural context and introduces additional complicated facets that may be associated to the context.

Many studies have focused on understanding the epistemic thinking of people within different social and cultural contexts (Nisbett et al. 2001; Tsai 1998; Weinstock 2010). As an essential example, different epistemologies exist in the prevalent manners of thinking among young adults from North American and Asian countries. Analytical thinking (following Greek philosophy) tends to be more prevalent within North American, whereas holistic thinking (following Confucian philosophy) is more common in Asian countries (Nisbett et al. 2001). The differences in the prevalent manners of thinking within these two contexts could have potentially influenced these students' personal styles of thinking as they have been exposed to both styles of thinking through their study experiences in both the Chinese and US educational systems.

11.1.1 Impact of This Research

As to the greater impact, this current research has expanded the understanding of the field of epistemological development in the following different aspects.

11.1.2 Applications of Perry's Theory Among Engineering Students

The domain specificity of epistemological models is a new direction in epistemological studies (Hofer 2001). The domain specificity of the epistemological theory has attracted attention in multiple areas, such as mathematics (Schoenfeld 1985) and science (Bell and Linn 2002). Here, the domain-specific application of Perry's intellectual and ethical developmental model among engineering students allows the exploration of these students' demonstrations of the thinking styles, as depicted in Perry's theory. The discipline-specific perspective of this research provides practical examples of the operationalization of the different thinking styles in Perry's theory drawn from an engineering context.

Most of the prior studies on engineering students either applied quantitative methods to explore their epistemological development (Choi et al. 2012) or used qualitative methods to evaluate the cognitive complexity of students by assigning scores of Perry's positions to the transcripts (Pavelich and Moore 1996; Wise et al. 2004). By using a mixed methods approach, this work has explored the manifestations of different thinking styles by collecting and analyzing qualitative data and has, therefore, produced a detailed list of various practical instances of these thinking styles. The focus on engineering students and their demonstrations of different thinking styles allows for a greater in-depth understanding of engineering students' epistemological development. Moreover, the demonstrations of these thinking styles among engineering doctoral students can possibly be utilized for the design and development of domain-specific assessment tools for epistemological development in engineering.

11.1.3   Extension of Perry's Theory

Most of the former studies on the applications of Perry's theory (Pavelich and Moore 1996; Wise et al. 2004; Zhang 1995, 1999, 2000, 2002; Zhang and Hood 1998; Zhang and Watkins 2001) focused on the undergraduate level students except in the instances of a few studies among liberal arts doctoral students, which then suggested the presence of a higher level of epistemological development among doctoral students (Kitchener and King 1981). No substantial studies have been conducted to explore doctoral students who are more likely to exhibit the higher levels thinking in Perry's theory. The higher levels of thinking, especially *Commitment within Relativism,* were rarely explored in depth in prior epistemological studies (Moore 2002).

This research, however, focuses on exploring the higher levels of Perry's scale (*Relativism* and *Commitment within Relativism*) among doctoral students. The emphasis on the higher levels of Perry's theory allows for the identification and operationalization of these stages of thinking. Some of the practical instances identified among the students matched the original findings within Perry's study. For example, according to Perry, the theme of discerning information is the core features of relativistic thinking. Also, the theme of taking major responsibilities is one the main characteristics of commitment to relativistic thinking. Both of these elements were identified among the population in study here.

In addition to the findings that are similar to Perry's study, some other practical instances identified among the students expanded the original findings in Perry's study by applying the theory among doctoral students. Therefore, these instances exhibit additional features that are particular to doctoral programs. For example, many students with relativistic thinking mentioned about the role of their advisors/professors as a guide while they had reserved their right to independently conduct research. Another example that is particular to doctoral programs is that some students with commitment to relativistic thinking spoke about their status of doing research as not depending on their advisors. Instead, they referred to themselves as the lead of their research. Essentially, they felt it was up to them to decide the direction of their research.

11.1.4 Understanding of Chinese Doctoral Students' Epistemological Development

Current available studies on the applications of Perry's theory among Chinese students (Zhang 1995, 1999, 2000, 2002; Zhang and Hood 1998; Zhang and Watkins 2001) focused on the undergraduate level students. Studies on Chinese students who were pursuing graduate degrees in the US indicated possible representations of complicated thinking among these Chinese students (Jiang 2010). However, no direct studies have been conducted to explore the Chinese engineering doctoral students' epistemological developmental status. The overall profile that has been hereby gathered from the quantitative survey results shows that the prominent thinking of a major portion of the Chinese engineering doctoral students is relativistic thinking or commitment to relativistic thinking. This finding may provide an essential snapshot for the epistemological development profile of Chinese engineering doctoral students within US graduate programs as a whole.

The research on the epistemological development of Chinese engineering doctoral students in the US can serve as an example for further investigations about the epistemological development of other foreign-born students. Considering the diverse graduate student populations within US institutions, the similar studies using theoretical and methodological frameworks proposed here can potentially benefit a larger student population by permitting a similar and deeper understanding of their epistemological development.

11.1.5 Explorations of Factors Associated with Relativistic Thinking

The high levels of thinking in Perry's theory, i.e., relativistic thinking and commitment to relativistic thinking, are regarded as resembling the thinking patterns of expert scientists and engineers (Prince and Felder 2006). The characteristics of these high levels of thinking in Perry's theory are also considered as parallels to those of *a deep approach* to learning (Felder and Brent 2004b). Students who often use a deep approach to learning tend to "take primary responsibility for their own learning and are perfectly comfortable challenging the assertions of authorities" (Felder and Brent 2004b, p. 3). Among the current literature on the preferable ways to promote this approach to learning, authentic problems were found to challenge students' thinking, motivate students' learning, and keep them engaged in the task, therefore promoting a deep approach to learning (Prince and Felder 2006).

Similar to these findings of ways to promote a deep approach to learning, the encounters with open-ended or real-life problems were found to be related to students' relativistic thinking in this research. In addition, a number of other factors were also identified in this research. The factors identified here, including the people around the students, the contextual and the experiential factors, all indicate potential ways to promote relativistic thinking. For example, many students spoke of the benefit of working on group projects or studying in a group as a means for helping them think more broadly and critically. These factors can provide practical suggestions and/or guidance for the design of instructions and learning with the intent to promote relativistic thinking and/or a deep approach to learning among students.

11.2 Future Directions

Multiple future research directions can be pursued to advance the current understanding of the field of epistemology and the applications of theories into educational practices. Some immediate measures can be taken to expand this current research, considering the limitations involved in the design and implementation processes.

First of all, the modification of Zhang's Cognitive Development Inventory has allowed for the measurement of the four styles of thinking in Perry's theory. One detailed position, however, Position 4b-*Relativism Subordinate* of Perry's theory, was only potentially reflected by a few items in the survey. Nonetheless, the qualitative research results from the students who were in the *Relativism* group seemed to reflect a true relativistic thinking where students evaluate information and evidence instead of thinking in a "relativistic" manner because this is what the authoritative figures (e.g., professors, or instructors) want, which is primarily the characteristic of thinking within Position 4b. Therefore, to further explore the differences between Position 4b-*Relativism Subordinate* and Position 5-*Relativism* and potentially construct a survey instrument that would better separate these positions, future studies can be designed to focus on those students who are in the early stage of their doctoral studies, master's students, or even undergraduates in their junior or senior years. These students are more likely to have a larger representation of students who would reflect both Position 4b and Position 5 when compared to doctoral students.

Second, the study participants were recruited from five universities in the US, but the number of participants can be further expanded for the researchers to control for the different factors that are related to their epistemological development. An increase in the number of participants can help to further the understanding of the impact of different factors on the students' epistemological development. This current research provides a basic understanding about some potential factors related to students' epistemological development, such as their academic progress and their places of origin. However, to have a more thorough understanding of these different factors, it is important to have a large and varied sample to allow for the control of some of the factors when one tries to understand the impact from some other factors.

In addition to the above-mentioned areas, some other future directions are proposed here to further investigate young adults' epistemological development.

As is mentioned in the discussion section, the characteristics of the high levels thinking styles in Perry's theory, i.e., relativistic thinking and commitment to relativistic thinking, are very similar to the characteristics of the deep approach to learning (Felder and Brent 2004b). Therefore, one potential direction by which to carry this research forward rests on the translation of the factors that are associated with relativistic thinking into ways to advance students' learning. Some of the factors that are associated with relativistic thinking and can potentially be helpful to students' development of a deep approach to learning include using group-based projects, engaging students in group study and group discussions, and so on. To translate these factors into students' learning may also imply the alterations of training objectives, the design of learning environments, and the design and implementation of instructional methods, many of which involve multiple stakeholders, including policy makers, administrators, professors, staff members, and students.

Also, it seemed that the factors of people, including students' major advisors, course instructors, lab mates, classmates, etc., play an important role in the shaping of students' epistemological thinking. Further research can be conducted to better understand the interactions between the students and these different groups. Research on these topics can be potentially drawn from the literature in areas such as the social network and interactions, interpersonal communications, and so on.

Additional research can also be performed in the area of assessment. The practical instances that were identified among the qualitative data operationalized different thinking styles in Perry's theory among engineering students. These practical instances can potentially serve as the basis for the development of a new assessment tool that is specific to engineering students' epistemological thinking.

Considering the large representation of foreign students within US doctoral programs, further research can be conducted to expand our current understanding of epistemological development to similarly include students from other ethnic groups in US institutions. The research conducted among Chinese engineering doctoral students can serve as an example to inform these further studies among students from other ethnic groups. The methodological design used here, an explanatory approach, can be employed so that researchers are enabled to both have an overall understanding of the studied students' epistemological developmental status using quantitative data and continue by uncovering details about the students' epistemic thinking using qualitative data. This approach or additional alternatives of quantitative and/or qualitative methods can be beneficial in the investigation of the epistemological thinking of students from multiple ethical groups.

Beyond further research conducted within US institutions, additional studies can also be designed to compare Chinese students from contexts other than US doctoral programs. As was noted in the Discussion section of this text, different epistemologies are present in the prevalent manners of thinking among young adults from North American and Asian countries (Nisbett et al. 2001). Therefore, it is interesting to design comparative studies to understand the epistemological development of Chinese students from both US and Chinese institutions. These studies can help to discover factors that are essential to one's epistemological development.

In summary, these multiple future directions and different research designs and methods can potentially help to advance our current understanding of young adults' epistemological development. Moreover, these future studies can translate theories and findings in epistemology into the context of higher education, especially into the context of engineering education. The translation of epistemological theories into practices shall be found useful when applied across varied areas, such as the design of training goals for students, the design and implementation of instructional methods, the interactions between advisors and students, and so forth.

11.3   Conclusion

In summary, this research provided *a picture of the epistemological development of Chinese engineering doctoral students in US institutions* within the context of Perry's theory by offering a first-hand understanding of the Chinese engineering doctoral students' epistemological developmental stages in both quantitative and qualitative manners. With the overall profile found through the survey and the practical instances of each thinking style identified in the qualitative data, this research was able to generate a fundamental understanding about these students' epistemological development stages within the context of Perry's theory for the first time.

This research added to the current body of knowledge regarding epistemology in several ways. It applies Perry's theory to a new population, i.e., the Chinese engineering doctoral students who are studying in US institutions. By applying the concepts of Perry's theory to this group, this work can serve as an example for similar explorations of epistemological development among other nationalities within the US institutions. Considering the large representation of Chinese doctoral students involved in the US engineering disciplines, potential findings from this research can also serve as a reference point for the decision-making process in the design of curriculum, the adoption of various pedagogical methods, and organizations of other educational activities.

Also, since this research focuses on doctoral students, it explored the higher levels of Perry's model (i.e., *Relativism* or *Commitment within Relativism*). Moreover, because of the emphasis on engineering students, this research has also explored the engineering specificity of Perry's theory, which allows the potential development of engineering-specific assessment tools for epistemological development. The exploration of different factors that are associated with the higher levels of Perry's theory expands the current understanding of the potential ways to adjust the design of teaching and learning to promote students' development toward relativistic thinking.

Multiple future directions and different research designs and methods can be followed to extend our current understanding of young adults' epistemological development. The translation of theories and findings in epistemology into the context of higher education, in particular the field of engineering education, can potentially facilitate the development toward more advanced and complex manners of thinking among the engineering students at both the collegiate and graduate levels.

References

Bell, P., & Linn, M. C. (2002). Beliefs about science: How does science instruction contribute? In B. K. Hofer & P. R. Pintrich (Eds.), *Personal epistemology: The psychology of beliefs about knowledge and knowing.* NJ: Erlbaum, Mahwah.

Choi, I., Hong, Y. C., Gattie, D. K., et al. (2012). Promoting second-year engineering students' epistemic beliefs and real-world problem-solving abilities through case-based E-learning resources. In *2012 Proceedings of the American Society for Engineering Education,* San Antonio, TX.

Cox, M. F., London, J. S., Ahn, B., Zhu, J., Torres-Ayala, A. T., Frazier, S., et al. (2011). Attributes of success for engineering Ph.D.s: perspectives from academia and industry. In *2011 Proceedings of the American Society for Engineering Education,* Vancouver, BC, Canada.

Felder, R. M., & Brent, R. (2004). The intellectual development of science and engineering students. 2. Teaching to promote growth. *Journal of Engineering Education, 93*(4), 279–291.

Gardner, S. K. (2008). What's too much and what's too little?: The process of becoming an independent researcher in doctoral education. *The Journal of Higher Education, 79*(3), 326–350.

Golde, C. M., & Walker, G. E. (2006). *Envisioning the future of doctoral education: Preparing stewards of the discipline.* Jossey-Bass-Carnegie Foundation for the Advancement of Teaching, San Francisco, CA: Carnegie Essays on the Doctorate.

Hofer, B. K. (2001). Personal epistemology research: Implications for learning and teaching. *Journal of Educational Psychology Review, 13,* 353–383.

Jiang, X. (2010). *Chinese engineering students' cross-cultural adaptation in graduate school.* Retrieved from ProQuest: The University of Michigan.

Kitchener, K. S., & King, P. M. (1981). Reflective judgment: Concepts of justification and their relationship to age and education. *Journal of Applied Developmental Psychology, 2,* 89–116.

Moore, W. S. (2002). Understanding learning in a postmodern world: reconsidering the Perry scheme of intellectual and ethical development. In B. Hofer & P. Pintrich (Eds.), *Personal epistemology: The psychology of beliefs about knowledge and knowing.* Mahwah, NJ: Lawrence Erlbaum Associates.

Nisbett, R. E., Peng, K., Choi, I., & Norenzayan, A. (2001). Culture and systems of thought: Holistic vs. analytic cognition. *Psychological Review, 108,* 291–310.

Pavelich, M. J., & Moore, W. S. (1996). Measuring the effect of experiential education using the Perry model. *Journal of Engineering Education, 85*(4), 287–292.

Prince, M. J., & Felder, R. M. (2006). Inductive teaching and learning methods: Definitions, comparisons, and research bases. *Journal of Engineering Education, 95*(2), 123–138.

Schoenfeld, A. H. (1985). *Mathematical problem solving.* San Diego, CA: Academic Press.

Tsai, C. C. (1998). An analysis of scientific epistemological beliefs and learning orientations of Taiwanese eighth graders. *Science Education., 82*(4), 473–489.

Weinstock, M. (2010). Epistemological development of Bedouins and Jews in Israel: Implications for self-authorship. In M. B. Baxter Magolda, E. G. Creamer & P. S. Meszaros (Eds.), *Development and assessment of self-authorship: Exploring the concept across cultures.* Sterling, VA: Stylus.

Wise, J., Lee, S. H., Litzinger, T. A., Marra, R. M., & Palmer, B. (2004). A report on a four-year longitudinal study of intellectual development of engineering undergraduates. *Journal of Adult Development, 11,* 103–110.

Zhang, L. F. (1995). *The construction of a Chinese language cognitive development inventory and its use in a cross-cultural study of the Perry Scheme.* Retrieved from ProQuest: The University of Iowa.

Zhang, L. F. (1999). A comparison of U.S. and Chinese university students' cognitive development: The cross-cultural applicability of Perry's theory. *The Journal of Psychology, 133*(4), 425–439.

Zhang, L. F. (2000). Are thinking styles and personality types related? *Educational Psychology, 20* (3), 271–283.

Zhang, L. F. (2002). Thinking styles and cognitive development. *The Journal of Genetic Psychology, 163*(2), 179–195.

Zhang, L. F., & Hood, A. B. (1998). Cognitive development of students in China and USA: opposite directions? *Psychological Reports, 82,* 1251–1263.

Zhang, L. F., & Watkins, D. (2001). Cognitive development and student approaches to learning: An investigation of Perry's theory with Chinese and U.S. university students. *Higher Education, 41,* 239–261.

Appendix A
Information Page to the Experts and the Expert Rating Form

© Springer Science+Business Media Singapore and Higher Education Press 2017
J. Zhu, *Understanding Chinese Engineering Doctoral Students in U.S. Institutions*,
East-West Crosscurrents in Higher Education, DOI 10.1007/978-981-10-1136-8

Information Page

Overview of Perry's Theory

Perry's theory (1970) describes nine positions along which college students' epistemological development takes place. The positions can be grouped into four major categories (Figure 1):

Dualism (Positions 1 and 2),

Multiplicity (Positions 3 and 4),

Relativism (Positions 5 and 6), and

Commitment (Positions 7, 8 and 9)

(Culver and Hackos,1982)

A brief description for each position in Perry's model is provided as follows. This study focuses on **Position 3, 4 and 5** (See the shaded box below).

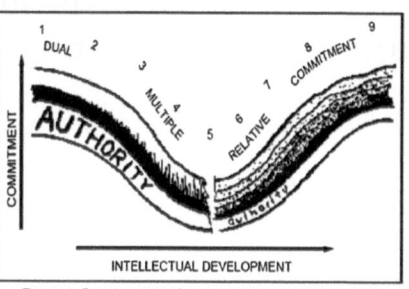

Figure 1. Perry's model (Culver and Hackos,1982, p. 223)

Dualism

Position 1 Basic Duality. A person in this position perceives any knowledge, act or value to be either "right" or "wrong". Any knowledge, act or value that differs from an Authority's world will be associated with error or evil, leaving the person with no alternative point to observe differently.

Position 2 Multiplicity Pre-legitimate. A person in this position perceives diversity in opinion and uncertainty as unnecessary confusion by poorly qualified Authorities, or a narrow area of freedom set by the Authority for student's own exploration.

Multiplicity

Position 3 Multiplicity Subordinate. A person in this position accepts uncertainty and diversity as unavoidable but only in areas where Authority hasn't attained the right answers. In areas where an Authority has no correct answers, the student is puzzled about the standards against which the Authority grades one's work. He supposes that an Authority grades based only on how well the answer is expressed.

Position 4 In this position, there are two alternative views:

4a. Multiplicity Correlate. In this position, Multiplicity is raised from a subordinate to a legitimate status, which is separate and equal to that of the Authority. Equal absolutism, i.e. "Everyone has a right to his own opinion" (Perry, 1970, p. 97) dominates the Multiplicity realm while in the Authority's domain there remains a right-wrong dualism.

4b. Relativism Subordinate. A person in this position perceives relativistic reasoning in terms of what the Authority wants. The student's weighing of different approaches to one problem and developing of his/her own thoughts occurs in the context of an Authority's realm out of a student's desire to confirm to Authority's demand.

Relativism

Position 5 Relativism. In this position, all knowledge and values, including the ones of authorities are perceived as relativistic (note: Authorities become authorities). A person here will analyze, weigh or evaluate different evidence, factors, or solutions to develop his/her own opinion, argument or solution.

Position 6 Commitment Foreseen. A person in this position realizes a necessity to commit oneself in a relativistic world. Acknowledging that reason itself has a limit and cannot fully justify or assure him, he must commit himself through his own faith, assuming the responsibilities associated with his/her choice.

Commitment

Position 7, 8 and 9. These positions describe different levels or scopes of commitment to a relativistic view. A person undertakes responsibilities in different major areas of life and experiences the implications of the commitment.

References:
Culver, R.S., and Hackos, J.T. (1982). Perry's model of intellectual development, Engineering Education, 72, 221–226.
Perry, W. G. (1970). Forms of intellectual and ethical development in the college years: A scheme. New York: Holt, Rinehart and Winston.

Page1/2

Directions: Please rate the degree to which the statement reflects the view of Perry's Positions 3, 4a, 4b, and 5 using the following scale:

Not at all --1
Only a little -- 2
Somewhat -- 3
Very much -- 4

Please refer to the shaded box on the **Overview of Perry's Theory** page for more information.

	Position 3 - Multiplicity Subordinate				Position 4a -Multiplicity Correlate				Position 4b - Relativism Subordinate				Position 5 - Relativism			
	1	2	3	4	1	2	3	4	1	2	3	4	1	2	3	4
e.g. Everyone has a right to his/her own opinion. Anything goes.	○	○	○	○	○	○	○	○	○	○	○	○	○	○	○	○
1 In areas where there is no one correct viewpoint but multiple viewpoints, I often feel unclear as to the standards that instructors use to evaluate students' viewpoints.	○	○	○	○	○	○	○	○	○	○	○	○	○	○	○	○
2 When experts in a particular field disagree with one another, no one really knows the answer.	○	○	○	○	○	○	○	○	○	○	○	○	○	○	○	○
3 Taking a stand during an academic debate requires reasoning and taking risks.	○	○	○	○	○	○	○	○	○	○	○	○	○	○	○	○
4 In issues where experts have no consensus, it can be confusing, because there is no way to prove whether one viewpoint is more reasonable than the other.	○	○	○	○	○	○	○	○	○	○	○	○	○	○	○	○
5 It is not difficult for me to give up ideas and opinions I hold if I find that my classmates' ideas sound more reasonable.	○	○	○	○	○	○	○	○	○	○	○	○	○	○	○	○
6 In an academic debate, there are merits in both sides' views, therefore, the criteria to decide which side wins are not clear.	○	○	○	○	○	○	○	○	○	○	○	○	○	○	○	○
7 I think it's unfair that instructors give a better grade to one student's answer than the other's answer, even when there is no clear answer for the problem.	○	○	○	○	○	○	○	○	○	○	○	○	○	○	○	○
8 When I solve a problem, I often think through several alternatives to find the best solution.	○	○	○	○	○	○	○	○	○	○	○	○	○	○	○	○
9 Since people's views are influenced by their educational backgrounds, almost any view could be as valid as any other.	○	○	○	○	○	○	○	○	○	○	○	○	○	○	○	○
10 Where authorities differ, a student's opinion should be graded only on how well it is expressed.	○	○	○	○	○	○	○	○	○	○	○	○	○	○	○	○
11 Integrating ideas is my favorite part of writing a paper.	○	○	○	○	○	○	○	○	○	○	○	○	○	○	○	○
12 I enjoy classes with a seminar format where students can exchange their ideas so I can critique my own perspectives on the subject matter.	○	○	○	○	○	○	○	○	○	○	○	○	○	○	○	○
13 If you really get deeper into the material in a particular field, what you find is that nobody understands it.	○	○	○	○	○	○	○	○	○	○	○	○	○	○	○	○
14 When writing an open-ended essay, I take a stance after thinking about other possible viewpoints.	○	○	○	○	○	○	○	○	○	○	○	○	○	○	○	○
15 I try to think in an independent manner, because thoughts that appear to be independent get good grades.	○	○	○	○	○	○	○	○	○	○	○	○	○	○	○	○
16 I prefer to be in a class that is loosely structured where the students take most of the responsibility for the structure of the class.	○	○	○	○	○	○	○	○	○	○	○	○	○	○	○	○
17 I enjoy a class where opportunities are provided for me to pull together connections among various subject areas and then construct an adequate argument.	○	○	○	○	○	○	○	○	○	○	○	○	○	○	○	○
18 It seems to me that some instructors try to get you to look at something in a complex way by weighing multiple factors at once.	○	○	○	○	○	○	○	○	○	○	○	○	○	○	○	○
19 I find I can detach myself emotionally from problems and look at their various sides in order to formulate a judgment.	○	○	○	○	○	○	○	○	○	○	○	○	○	○	○	○
20 I would like opportunities to think on my own and make connections between the issues discussed in class and in other areas I'm studying.	○	○	○	○	○	○	○	○	○	○	○	○	○	○	○	○

Page2/2

Directions: *Please rate the degree to which the statement reflects the view of Perry's Positions 3, 4a, 4b, and 5 using the following scale:*

Not at all --1
Only a little -- 2
Somewhat -- 3
Very much -- 4

Please refer to the shaded box on the **Overview of Perry's Theory** page for more information.

	Position 3-Multiplicity Subordinate				Position 4a -Multiplicity Correlate				Position 4b - Relativism Subordinate				Position 5 - Relativism			
	1	2	3	4	1	2	3	4	1	2	3	4	1	2	3	4
21 It doesn't seem fair when grades aren't proportional to work efforts. Many times a person can receive a better grade on a paper that he hasn't worked hard on than a paper that he really worked on.	○	○	○	○	○	○	○	○	○	○	○	○	○	○	○	○
22 It is very hard for me to accept a teacher's view on controversial issues when he/she does not provide enough evidence to support his/her view.	○	○	○	○	○	○	○	○	○	○	○	○	○	○	○	○
23 I am certain of one thing- even if there is an absolute truth, we will never know about it and, therefore, there is no correct answer to most questions.	○	○	○	○	○	○	○	○	○	○	○	○	○	○	○	○
24 I enjoy working with complex ideas in which experts have no consensus.	○	○	○	○	○	○	○	○	○	○	○	○	○	○	○	○

Do you have any additional comments or suggestions in regards to the survey in general or to specific items that you just rated?

Please rate your level of engagement with Perry's theory. Please check all that are applicable to you.

☐ I have been engaged in research efforts using Perry's theory.

☐ I have taught some class related to Perry's theory.

☐ I have attended some training sessions (classes, workshops, etc.) related to Perry's theory.

☐ Other, please specify --

Appendix B
Complete Survey with the Modified ZCDI and the KCM Subscale

Page 1/3
Directions:

The following survey contains different statements about your views in topics like, criteria for knowledge, process of learning, role of teachers and learners, etc. Please rate the degree to which you agree/disagree with the following statements. Please rate all the statements including the ones that may not be applicable to you.

Strongly Disagree -- SD
Disagree -- D
Neutral--N
Agree -- A
Strongly Agree -- SA

	The degree to which I agree with the statement--				
	SD	D	N	A	SA
In answering essay questions from homework, I would be better off using notes from the teacher's lecture than materials I read on my own.	○	○	○	○	○
In any debate, it is almost always true that one party's viewpoint is correct and the other's is wrong.	○	○	○	○	○
The true experts in a particular field generally agree with one another.	○	○	○	○	○
It doesn't seem fair when grades aren't proportional to work efforts. Many times a person can receive a better grade on a paper that he hasn't worked hard on than a paper that he really worked on.	○	○	○	○	○
Among the most important goals in my life, I have decided the one to be achieved first.	○	○	○	○	○
It is very hard for me to accept a teacher's view on controversial issues when he/she does not provide enough evidence to support his/her view.	○	○	○	○	○
Nowadays, I spend more time trying to accomplish my personal goals than I spend thinking about them.	○	○	○	○	○
I would like opportunities to think on my own and make connections between the issues discussed in class and in other areas I'm studying.	○	○	○	○	○
In an academic debate, there are merits in both sides' views, therefore, the criteria to decide which side wins are not clear.	○	○	○	○	○
I prefer classes with a seminar format where students can exchange their ideas so I can critique my own perspectives on the subject matter.	○	○	○	○	○
Essay questions may appear to have more than one answer, but there is only one right answer to any question.	○	○	○	○	○
Since people's views are influenced by their educational backgrounds, almost any view could be as valid as any other.	○	○	○	○	○
Taking a stand during an academic debate requires reasoning and taking risks.	○	○	○	○	○
Good teachers never let you leave the classroom with doubts about the subject matter.	○	○	○	○	○
Once I set a goal, I never give up pursuing it, no matter under what circumstances.	○	○	○	○	○
I schedule activities in my life according to the long-term goals I have set.	○	○	○	○	○
It seems to me that some instructors try to get you to look at something in a complex way by weighing multiple factors at once.	○	○	○	○	○
I prefer to be in a class where the instructor involves the students to contribute to the structure of the class.	○	○	○	○	○
I never work on anything that does not guarantee success.	○	○	○	○	○
I am not satisfied with anything below an "A" in a course.	○	○	○	○	○
A teacher's only important job is to communicate the facts of his or her field to students.	○	○	○	○	○
I have set priorities on my activities so that I could achieve my major goals.	○	○	○	○	○
We should always allow our reason to govern our behaviors, not our emotion.	○	○	○	○	○
When confronted with a controversial issue, it is wiser to side with the teacher than getting involved with some endless debate.	○	○	○	○	○

© Springer Science+Business Media Singapore and Higher Education Press 2017
J. Zhu, *Understanding Chinese Engineering Doctoral Students in U.S. Institutions*,
East-West Crosscurrents in Higher Education, DOI 10.1007/978-981-10-1136-8

Directions:
The following survey contains different statements about your views in topics like, criteria for knowledge, process of learning, role of teachers and learners, etc. Please rate the degree to which you agree/disagree with the following statements. Please rate all the statements including the ones that may not be applicable to you.

Strongly Disagree -- SD
Disagree -- D
Neutral--N
Agree -- A
Strongly Agree -- SA

	The degree to which I agree with the statement--				
	SD	D	N	A	SA
I find I can detach myself emotionally from problems and look at their various sides in order to formulate a judgment.	○	○	○	○	○
I find it helpful to be in a class where opportunities are provided for me to pull together connections among various subject areas and then construct an adequate argument.	○	○	○	○	○
It is not difficult for me to give up ideas and opinions I hold if I find that my classmates' ideas sound more reasonable.	○	○	○	○	○
Where authorities differ, a student's opinion should be graded only on how well it is expressed.	○	○	○	○	○
I think it's unfair that instructors give a better grade to one student's answer than the other's answer, even when there is no clear answer for the problem.	○	○	○	○	○
I often consider the potential effects of my behavior on the good of the society.	○	○	○	○	○
I prefer tasks dealing with a single, concrete problem, rather than multiple, or general ones.	○	○	○	○	○
In areas where there is no one correct viewpoint but multiple viewpoints, I often feel unclear as to the standards that instructors use to evaluate students' viewpoints.	○	○	○	○	○
I have assumed major responsibilities in several areas of my life.	○	○	○	○	○
Disagreements regarding important issues should be left to the experts.	○	○	○	○	○
If you really get deeper into the material in a particular field, what you find is that nobody understands it.	○	○	○	○	○
The key to understanding a course is learning to think the way the teacher wants you to think.	○	○	○	○	○
Teachers should give you the right answers when you cannot find them on your own.	○	○	○	○	○
When experts in a particular field disagree with one another, no one really knows the answer.	○	○	○	○	○
Teachers should make sure that students produce the right answers to any questions.	○	○	○	○	○
As a student, if I do not get good grades, I would be almost good for nothing.	○	○	○	○	○
I'd rather be given a definite topic for a paper than having a lot of topics to choose from.	○	○	○	○	○
When a teacher takes a particular position on a controversial issue, you can be almost certain that he/she is on the right side.	○	○	○	○	○
What happens in my life is usually up to me.	○	○	○	○	○
When writing an open-ended essay, I take a stance after thinking about other possible viewpoints.	○	○	○	○	○
Once a person gets all the facts, he/she would find out that there is only one right answer to a question.	○	○	○	○	○
In issues where experts have no consensus, it can be confusing, because there is no way to prove whether one viewpoint is more reasonable than the other.	○	○	○	○	○
Integrating ideas is my favorite part of writing a paper.	○	○	○	○	○
When I solve a problem, I often think through several alternatives to find the best solution.	○	○	○	○	○

Page 3/3
Directions:
The following survey contains different statements about your views in topics like, criteria for knowledge, proces of learning, role of teachers and learners, etc. Please rate the degree to which you agree/disagree with the following statements. Please rate all the statements including the ones that may not be applicable to you.

Strongly Disagree -- SD
Disagree -- D
Neutral--N
Agree -- A
Strongly Agree -- SA

	The degree to which I agree with the statement--				
	SD	D	N	A	SA
The only thing that is certain is uncertainty itself.	○	○	○	○	○
Forming your own ideas is more important than learning what the textbooks say.	○	○	○	○	○
A really good way to understand a textbook is to reorganize the information according to your own personal scheme.	○	○	○	○	○
You should evaluate the accuracy of information in textbooks if you are familiar with the topic.	○	○	○	○	○
Wisdom is not knowing the answers, but knowing how to find the answers.	○	○	○	○	○
Today's facts may be tomorrow's fiction.	○	○	○	○	○
The most important part of scientific work is original thinking.	○	○	○	○	○
Even advice from experts should be questioned.	○	○	○	○	○
I try my best to combine information across chapters or even across classes.	○	○	○	○	○
A sentence has little meaning unless you know the situation in which it was spoken.	○	○	○	○	○
I find it refreshing to think about issues that experts cannot agree on.	○	○	○	○	○

Appendix C
Demographic Survey

What is your gender?

- Male
- Female

What is your primary language (i.e., the one you speak most of the time)?

- Chinese
- English
- Other, please specify

What is your current marital status?

- Single
- Married
- Widowed
- In a committed relationship
- Rather not say

How many children under 16 years old live in your household?

- None
- 1
- 2
- 3 or more

How old are you?

- < 22
- 22-25
- 25-30
- 30-35
- > 35

Of which province (or region) of China are you from?

© Springer Science+Business Media Singapore and Higher Education Press 2017
J. Zhu, *Understanding Chinese Engineering Doctoral Students in U.S. Institutions*,
East-West Crosscurrents in Higher Education, DOI 10.1007/978-981-10-1136-8

Which of the following best describes the area you grew up (before college)?

○ urban

○ suburban

○ rural

What is your religion?

○ Atheism

○ Buddism

○ Catholicism

○ Christianity

○ Islam

○ Taoism

○ Rather not say

○ Other, please specify

What is the highest educational attainment of your father?

○ Less than secondary school

○ Secondary school/high school

○ Some college

○ Bachelor's degree

○ Master's Degree

○ Doctoral Degree

○ Professional Degree (e.g. MD, JD)

○ Other, please specify

What is the highest educational attainment of your mother?

○ Less than secondary school

○ Secondary school/high school

○ Some college

○ Bachelor's degree

○ Master's Degree

○ Doctoral Degree

○ Professional Degree (e.g. MD, JD)

○ Other, please specify

Please provide the information about your education (starting from your bachelor's degree up till now).

	Name of Institution	Degree type (B.S., M.S., etc.)	Started date (MM/YYYY)	Finished date / Expected date of degree completion (MM/YYYY)	Major
1					
2					
3					
4					
5					
6					

Which of the following best describe you in your current graduate study? (Check all that apply)

☐ I am working on my coursework

☐ I have completed all of my course work

☐ I have passed the Qualification Exam

☐ I have passed the Preliminary Exam

☐ I am in my dissertation stage

☐ I have defended my dissertation

How long have you worked prior to your current graduate study?

○ I have no prior work experience

○ Less than 1 year

○ 1-3 years

○ 3-5 years

○ more than 5 years

If you have prior work experiences (as in the last question), what kind of organizations or sectors did you work for? (Check all that apply). If you have no prior work experience, please skip this question.

☐ National or local governmental sector

☐ Educational intitutions (e.g. universities, middle schools, etc.)

☐ For-profit private sector (e.g. corporations, small business, etc.)

☐ Not-for-profit sector

☐ Self-employed

☐ Other, please specify

Appendix D
Interview Protocol

Interview Protocol访谈问题

Introduction: Thank you for your willingness to participate in today's interview. The goal of this research is to understand Chinese engineering students' perspective toward learning and their cognitive development.

开场白:谢谢你来参与今天的访谈,这个研究主要是想了解工程专业的中国学生对学习的认知和学生的认知发展。

1. Introductory Question in Perry's original work Perry的研究中的访谈问题

Please take a few minutes and think about the past year (or semester for year-1 students). What stands out in your learning experience? (Change the past year into "since your doctoral study" and "since your undergraduate study")

你能不能先用几分钟想一下过去一年中(或者一个学期,对第一年的学生来说),在你的学习里,有什么特别让你难忘/印象深刻的经历呢?(从你开始念Ph.D.开始、从本科到现在)

2. Role of Learning学习的作用

What about your perspectives on the value of the things you have learned in the past year. What things have you learned that you think are important? Why? (Change the past year into "since your doctoral study" and "since your undergraduate study")

我们来看看你对过去一年中学到的东西的看法?你学到了什么对你来说特别重要的东西呢?你为什么这么觉得呢?(从你开始念Ph.D.开始、从本科到现在)

© Springer Science+Business Media Singapore and Higher Education Press 2017   177
J. Zhu, *Understanding Chinese Engineering Doctoral Students in U.S. Institutions*,
East-West Crosscurrents in Higher Education, DOI 10.1007/978-981-10-1136-8

3. Role of Learner学习者的角色

    a. Now, think about yourself as a learner in the classroom, in a research group, or in a project team. What role do you play, what method do you use, to make learning more effective for you?

    现在想一下,作为一个学生,在你修课、或者在你做研究的过程中,你觉得自己是扮演什么样的角色来学习的呢?

    b. How did this role come about?

    这样的角色是怎么形成的?

4. Role of Instructors老师的角色

    a. As you think about your instructors, professors, advisor(s), what role do you think they have played that made you learn effectively?

    现在想一下,你的教授、导师们,你觉得他们在你的学习中,起了什么作用,或者扮演了什么角色,是对你的学习很有帮助的?

    b. How did you realize that these roles/functions are useful for your learning?

    你是如何认识到这样的角色/作用是对你比较有帮助的?

5. Role of Peers同学的角色

    a. What about other students in your classes, your research team? What kinds of interactions with them would help you with your learning?

    现在想一下,你班上的其他学生,你的研究小组里的其他成员,你和他们之间什么样互动是有助于你的学习的?你觉得他们在你的学习中起了什么作用,或者说扮演了什么样的角色?

    b. How did you realize that these roles/functions/interactions are useful for your learning?

    你是如何认识到这样的角色/作用是对你比较有帮助的?

6. Perception of the Evaluation of Their Work 对自己工作的评估

    a. As you think about the work you have done in your classes or research, what kinds of feedback/suggestions do you think are helpful to your learning?

    现在想一下,你在课上作业,或者project,或者在研究中所作的工作,你觉得自己得到什么类型的反馈或者意见,能够更有助于你的学习呢?

b. How did you realize that these feedback or suggestions can help your learning?

你是怎么认识到这些类型的反馈能有助于你的学习呢?

7. Varying Points of View面对不同的观点

In your learning, have you experienced varying points of view?

在你的学习经历中,你有碰到过别人和你观点不同的情况吗?

If the response from the interviewee is "yes":

如果回答是"有"的话:

a. Could you please describe the situation in which you encountered the varying points of view?

你可不可以描述一下都是在那些情况下,你遇到这些不同的观点?

b. How do you decide what to accept or believe? Why?

这种情况下,你是如何做决定的呢?为什么?

If the response from the interviewee is "no":

如果回答是"没有"的话:

a' If you do, how would you react or decide what to accept or believe? Why?

你可不可以假想一下,如果你碰到的话,你会如何做决定呢?为什么?

8. Role of the Environment 成长环境的作用

What kinds of factors influenced your learning when you grew up? Probing question

在你的成长过程中,你觉得有哪些因素对你的学习有比较重要的影响呢?

9. Educational Decision Making学习上的决定

Did you made any educational decisions last year that are important to you? Could you please describe the process of making these decisions? (Change the past year into "since your doctoral study" and "since your undergraduate study")

过去一年中(从你开始念Ph.D.开始、从本科到现在)在学业上你做过什么对你来说很重要的决定吗?你可不可以描述一下你作这些决定的过程?

10. (Only for the groups of *Relativism* and *Commitment*) Decision Making in other Areas (仅适用于在*Relativism*和*Commitment*小组的人) 其它方面的决定

Did you make any decisions last year that are important to you in areas other than the area of study? Could you please describe the process of making these decisions? (Change the past year into "since your doctoral study" and "since your under-graduate study")

过去一年中(从你开始念Ph.D.开始、从本科到现在)在学习以外的生活其他方面你做过什么对你来说很重要的决定吗?你可不可以描述一下你作这些决定的过程?

11. Open-ended question

Concerning our topic today, do you have anything else to add? (Prompt, i.e., that can help me understand your perspective toward your study?)

关于我们今天所谈论的话题,你还有什么其他要补充的吗?

Demographic questions 基本情况:

1. What year are you in your doctoral study? 你现在是博士第几年?
2. What stage are you at in your doctoral study? 你现在在博士什么阶段? (Course work, qualifying examination, preliminary, dissertation)
3. What major are you in? 你是什么专业?
4. What university are you in? 你在那个学校?
5. Where did you do your undergraduate degree? 你在哪里念的本科?
6. Did you have a master's degree or not before your doctoral study? 你在读博士之前念过硕士吗?
7. Where did you do your master's degree? 你在哪里念的硕士?

Thank you for joining in today's interview. That is all I have for today.

谢谢你参加今天的访问.访问到这里就结束了.

Appendix E
Dualistic Thinking-Demonstration Codes

Third level code list	Counts
1. Professors as the mediator of the right answers to questions or procedures	13
2. To learn the procedural knowledge/skills in doing research	7
3. To take in information in a passive manner in course work	4
4. Authoritative figure is very knowledgeable or capable, higher than the students	3
5. Diversity in opinion has been given a place	3
6. To take order in a passive manner	2
7. To simply do the work without affecting the direction of the research	1
8. Uncertainty has been given a place	1

© Springer Science+Business Media Singapore and Higher Education Press 2017
J. Zhu, *Understanding Chinese Engineering Doctoral Students in U.S. Institutions*,
East-West Crosscurrents in Higher Education, DOI 10.1007/978-981-10-1136-8

Appendix F
Multiplistic Thinking-Demonstration Codes

Third level code list	Counts
1. To be aware of other students/faculty/researchers' opinions, ideas, progress, or other information WITHOUT discernment	5
2. To collect information in a large amount or in an active manner WITHOUT explicit discernment	3
3. To be aware of what is going on in the field (with NO explicit intent to inform one's own research or study)	2
4. To discover the limitation of authority	2
5. To accept the legitimacy of diversity in opinions	1
6. To accept the legitimacy of the uncertainty of knowledge	1
7. To understand the evaluation of process in terms of the length of a paper/dissertation	1

© Springer Science+Business Media Singapore and Higher Education Press 2017
J. Zhu, *Understanding Chinese Engineering Doctoral Students in U.S. Institutions*,
East-West Crosscurrents in Higher Education, DOI 10.1007/978-981-10-1136-8

Appendix G
Relativistic Thinking-Demonstration Codes

Third level code list	Counts
1. To regard professors/advisors' roles as a guide	50
2. To actively ask others questions or seek out help	38
3. To collect information in a large amount or in an active manner WITH explicit intent of discernment	36
4. To appreciate feedback from others	33
5. To find and/or solve problems	32
6. To discern information (e.g., papers, lectures, reports, etc.)	28
7. To take and/or test the feedback from others	26
8. To learn from others (generic)	23
9. To learn in an independent manner	23
10. To conduct or to anticipate conducting research in an independent manner	22
11. To defend one's positions/ideas and/or convince others	20
12. To think in an independent manner	19
13. To collaborate with others/to have teamwork skills	17
14. To explore and test the unknown	14
15. To master the knowledge at a deep level	14
16. To actively express one's idea (oral)	13
17. To appreciate conflicts and/or the diversity of opinions	12
18. To be aware of what is going on in the field (with an explicit intent to inform one's own research or study)	12
19. To expand one's scope of knowledge (with the intention to inform one's research/study)	9
20. To be aware of other students/faculty/researchers/experts' opinions, ideas, progress, or other information with discernment	7

(continued)

© Springer Science+Business Media Singapore and Higher Education Press 2017
J. Zhu, *Understanding Chinese Engineering Doctoral Students in U.S. Institutions*,
East-West Crosscurrents in Higher Education, DOI 10.1007/978-981-10-1136-8

(continued)

Third level code list	Counts
21. To communicate with people from diverse backgrounds	7
22. To appreciate the deep thinking/philosophy of professors/advisor	6
23. To learn to compromise	6
24. To produce sound research results	6
25. To appreciate the benefits of doing group work/projects	4
26. To appreciate the evaluative methods based on the quality of reasoning	4
27. To be innovative/to generate new ideas	4
28. To have a collaborative relations with professors	4
29. To provide feedback to others	4
30. To think outside the box	4
31. To actively seek validation from the authoritative figure of the field	3
32. To address real-life problems (technical)	3
33. To be aware of a broad picture of a research field	3
34. To enjoy the process of being critiqued	3
35. To try to take some level of leadership in research	3
36. To understand one's own technical/career interest	3
37. To understand/examine/check the ability of the advisor	3
38. To accumulate experiences in doing research in an active manner or a fast speed	2
39. To consider non-technical factors when doing technical research	2
40. To investigate/analyze (technical content) thoroughly before making a technical decision	2
41. To achieve agreement in the fundamentals	1
42. To be aware of the evaluative methods based on the quality of reasoning	1
43. To connect different pieces of information	1
44. To develop a systematic method of doing research/learning	1
45. To know the ability of the advisor in a given field	1
46. To present one's idea well in writing manner	1
47. To realize the need to balance between personal interest and other factors	1
48. To recognize/accept the relativistic nature of ideas/knowledge	1
49. To understand one's own technical ability	1

Appendix H
Commitment-Demonstration Codes

Third level code list	Counts
1. To make a decision balancing different tensions	26
2. To take the major responsibilities in research or learning	14
3. To understand the uncertainty/limitation/implications of making decisions	13
4. To have a clear goal	12
5. To be the lead of oneself (life, goal, research, learning)	9
6. Anticipation of being the lead in one's career, research, work, etc.	8
7. Stylistic attributes (to do practical research vs. to do basic research)	7
8. To prepare for future career	7
9. To be persistent despite uncertainty	6
10. To do life planning	6
11. To not depend on the advisor or instructors	6
12. To recognize the need to prioritize	6
13. Stylistic attributes (learning/doing research out of real interest vs. learning/doing research for the sake of doing it)	5
14. To be brave/to overcome challenges	5
15. To be self-motivated	5
16. To have a positive attitude	5
17. To have a strong motivation	5
18. Stylistic attributes (to be social professionally vs. to be isolated)	4
19. Stylistic attributes (to live a rich life vs. to live by essentials)	4
20. To commit to a certain type of faith	4
21. To transform one's way of life because of his/her faith	4
22. Stylistic attributes (to challenge oneself vs. to choose the easy way out)	3
23. To be devoted to the work	3
24. Anticipation of a strong motivation	2
25. Courage in spite of uncertainty	2
26. Stylistic attributes (focusing on self vs. focusing on others)	2
27. Stylistic attributes (keeping the research relevant, practical, constructive vs. using topics that are irrelevant)	2

(continued)

© Springer Science+Business Media Singapore and Higher Education Press 2017
J. Zhu, *Understanding Chinese Engineering Doctoral Students in U.S. Institutions*,
East-West Crosscurrents in Higher Education, DOI 10.1007/978-981-10-1136-8

(continued)

Third level code list	Counts
28. Stylistic attributes (to be social vs. to be timid)	2
29. Stylistic attributes (following real interest/passion in research vs. treating research as a job to earn a living)	2
30. To be independent in one's spirit, mentally or psychologically	2
31. To be professional	2
32. To commit to a stable relationship with understanding of its implications	2
33. To experience, and/or understand, and/or accept the limitation of reasoning	2
34. To set up and/or follow through a reasonable plan	2
35. To transform one's goals in life because of his/her faith	2
36. Anticipation of commitment in a certain faith	1
37. Anticipation of devotion to work	1
38. Anticipation of self-motivation	1
39. Concerned about success as a human being	1
40. Stylistic attributes (being patient vs. being hasty)	1
41. Stylistic attributes (Being prudent vs. Being impulsive)	1
42. Stylistic attributes (being wise vs. being emotional)	1
43. Stylistic attributes (to learn from industry vs. to stay in school)	1
44. Stylistic attributes (to gain skills vs. to earn a certificate)	1
45. Stylistic attributes (the status of adults vs. the status of children)	1
46. Stylistic attributes (to help others vs. to focus on oneself)	1
47. Stylistic attributes (to separate life and work vs. to put life and work together)	1
48. To be independent financially	1
49. To choose appropriate methods in relationships	1
50. To commit to a certain social/cultural group	1
51. To experience personal growth via a major decision-making process	1
52. To have an awareness of social responsibilities	1
53. To increase one's public visibility	1
54. To perceive technical research through the lens of inner faith	1
55. To summarize the trick of the trade	1

Appendix I
Relativistic Thinking-Factor Codes

Third level code list	Counts
1. Discussions with other students	33
2. Advisors	32
3. Senior lab mates or schoolmates	28
4. Course instructors	21
5. Failures, pressures, or obstacles	17
6. Being in the US	14
7. Exposures to course work in the US	14
8. Being in a US graduate school or university	11
9. Experiences with group project	11
10. Positive or negative role models	11
11. Research group	11
12. Encounters with open-ended problems/projects	8
13. Experiences with oral presentations	6
14. Parents	6
15. Professional conferences	6
16. Experiences of studying in a group	5
17. Learning to collect and/or read academic papers	5
18. Being challenged in one's thinking	4
19. Being in a competitive environment	4
20. Being in a doctoral program	4
21. Other professors (not one's advisor or course instructors)	4
22. Choosing a major, field, or area of interest	3
23. Current university	3
24. Discussion with others (generic)	3
25. Experiences with technical content	3
26. Experiences with writing reports	3
27. Internship experiences	3
28. Media products (TV, movie, books, etc.)	3

(continued)

© Springer Science+Business Media Singapore and Higher Education Press 2017
J. Zhu, *Understanding Chinese Engineering Doctoral Students in U.S. Institutions*,
East-West Crosscurrents in Higher Education, DOI 10.1007/978-981-10-1136-8

(continued)

Third level code list	Counts
29. Students from other countries/backgrounds	3
30. Being in an environment without supervision	2
31. College experiences	2
32. Experiences during master's study	2
33. Experiences in some competitive tests	2
34. Industrial representatives or working professionals	2
35. Milestone examinations	2
36. Successes	2
37. Work experiences	2
38. Choosing courses of interest	1
39. Environment of a big city	1
40. High expectation	1
41. Other family members besides parents	1
42. Professional development opportunities or lectures	1
43. Researchers from other university	1
44. Spiritual or psychological needs	1
45. Spouses/partners of a committed relationship	1